The Chief Officer:
A Symbol Is a Promise

The Chief Officer:
A Symbol Is a Promise

Randy R. Bruegman

PEARSON

Prentice
Hall

Upper Saddle River, New Jersey

Library of Congress Cataloging-in-Publication Data

Bruegman, Randy R.
 The chief officer: a symbol is a promise/Randy R. Bruegman.—1st
ed.
    p. cm.
 Includes bibliographical references and index.
 ISBN 0-13-112501-X
 1. Fire departments—Management. 2. Leadership. I. Title.

 TH9158.B773 2004
 363.37'068—dc22

                          2004003247

Publisher: *Julie Levin Alexander*
Publisher's Assistant: *Regina Bruno*
Senior Acquisitions Editor: *Katrin Beacom*
Editorial Assistant: *Kierra Kashickey*
Senior Marketing Manager: *Katrin Beacom*
Channel Marketing Manager: *Rachele Strober*
Director of Production and Manufacturing: *Bruce Johnson*
Managing Editor for Production: *Patrick Walsh*
Production Liaison: *Julie Li*

Production Editor: *Penny Walker, The GTS Companies/York, PA Campus*
Manufacturing Buyer: *Pat Brown*
Creative Director: *Cheryl Asherman*
Senior Design Coordinator: *Christopher Weigand*
Cover Designer: *Christopher Weigand*
Cover Photo: *Randy R. Bruegman*
Composition: *The GTS Companies/York, PA Campus*
Printing and Binding: *R. R. Donnelley & Sons*
Cover Printer: *Phoenix Color Corporation*

Pearson Education Ltd.
Pearson Education Singapore, Pte. Ltd.
Pearson Education, Canada, Ltd.
Pearson Education—Japan

Pearson Education Australia PTY, Limited
Pearson Education North Asia Ltd
Pearson Educación de Mexico, S.A. de C.V.
Pearson Education Malaysia, Pte. Ltd.

10 9 8 7 6 5 4 3 2 1
ISBN 0-13-112501-X

CONTENTS

REVIEWERS

Kevin S. Brame M.A.
Battalion Chief
Orange County, California

J. Robert Brown, Jr.
Chief of Staff, Boone County Fire District
Columbia, Missouri

John M. Buckman III
Fire Chief, German Township Volunteer Fire Department
Evansvile, Indiana

Steven T. Edwards
Director, Maryland Fire and Rescue Institute of the University of Maryland
College Park, Maryland

Bruce Evans, M.P.A.
Fire Program Coordinator, Community College of Southern Nevada
Fire Captain, City of Henderson
Henderson, Nevada

Richard Johnson
Ohio Fire Academy
Office of State Fire Marshal
Reynoldsburg, Ohio

Richard Marinucci
Fire Chief
Farmington Hills, Michigan

Mario H. Treviño
Former Chief of Department, San Francisco Fire Department
San Francisco, California

PREFACE

Make a Difference Every Day!

It was about 4 p.m. in late September 2001 when my secretary asked me to go to the lobby of our office because there was someone who wished to see me. There stood a boy, eight or nine years old, with his mother, holding a big pickle jar full of dollar bills and coins. The jar must have weighed thirty pounds. It was all he could do to carry it, but there he stood, grinning from ear to ear, yet looking determined at the same time. He asked if I was the fire chief, and as I introduced myself, I noticed he was wearing a 9-1-1 pin with the saying "We will never forget." After our introductions, his mother told me that he was moved by the September 11 events and that he wanted to do something for the families of the firefighters who were lost that day. He took his allowance, went to the local Wal-Mart, and bought as many candy bars as he could. He then went from door to door in his neighborhood, selling them for a dollar each to raise money for the firefighters' families. He turned in almost $300. It was an amazing moment, but as quickly as this little boy had come, he was gone. No fanfare. He wanted nothing in return; he just wanted to help. He made a difference.

What motivated this boy to get involved and take action is the same characteristic—a drive, a sense of something bigger than oneself—that we see in the fire service every day. This book is about the spirit and commitment that young man displayed in putting service over self—the essence of the fire service.

When I began to outline this book, there were several ideas that I wanted to convey. The first was a review of the history and rich traditions of the fire service. It is a history that has helped to create a unique, shared value system that has driven our industry for hundreds of years. Our commitment to service, our willingness to sacrifice ourselves for others, is rooted in the origins of the fire service. It is a powerful culture that dates back over a thousand years and is seen every day in the actions of our personnel. I don't want us to lose our connection to that history, as it is the basis for what we are today.

This book is also about leadership, specifically about becoming a chief fire officer. The health of our industry will be determined by those who choose, and have the ability, to lead. This is very important. Throughout this book, I have attempted to share my perspective, my experiences, and the knowledge of others to help you answer your own questions. Is a chief fire officer's job the right one for you? As you move up in rank through the fire service, each step involves greater responsibility and expectations. Each step also requires a different set of skills. I have tried to focus on the preparations you will need for each step. I also provide suggestions for developing your own plan for success. This game plan begins by asking yourself where you wish to end your career. Once you've identified your target, your game plan will help you to achieve it. I have also tried to describe my own experiences to help you understand the complexities of the role of chief fire officer. Whether you aspire to be a battalion commander or the fire chief, the gold badge represents the end of past responsibilities and the beginning of many new ones. Some people have difficulty making the transition. In this book, I've attempted to provide help in making that transition successful both professionally and personally.

Finally, I focus on taking command from both a personal and a professional perspective. Taking over a new organization, building a team, and understanding the future impacts on the fire service is critical for anyone in a leadership position. As we move through each of these topics, the elements of change and organizational culture are explored, as each is essential in leading an organization today.

Like you, I have been influenced by several people during my career, as noted throughout this book. Ron Coleman, California State Fire Marshal, International Association of Fire Chiefs president (1989–1990), Fire Chief in Fullerton and San Clemente, California (the list is endless), has been my mentor and friend for almost twenty years. Another friend is Ray Picard, retired fire chief, Huntington Beach, California, whom I worked with for over a decade, on the Fire Service Accreditation Project. Ray is one of those people who has forgotten more than most of us will ever know. He is a brilliant man, far ahead of his time. Another mentor is Pete Burchard, city manager of Naperville, Illinois. I had the opportunity to work with Pete in Hoffman Estates, Illinois, and learned from him how to meld leadership into daily management. He is a master of that skill. Then there's my family—my wife, Susan, and my children, Josh, Stephanie, and Chris. When you read about the relocation blues in Chapter 4, think of them. Much of that account is about their experience.

But, most of all, this book is for those who are thinking about becoming a chief fire officer. It is about the fire service's commitment to make a difference every day. It is about the symbolism of the fire service that translates into meaningful ideals by which we live—courage, integrity, the work ethic, empathy, and a commitment to service over self. This book is dedicated to all in the fire service who, on a daily basis, risk their own lives in their service to others.

They understand that a symbol is a promise.

CHAPTER 1

Pride, Service,
Commitment

Without belittling the courage with which men have died, we should not forget those acts of courage with which men . . . have lived. The courage of life is often a less dramatic spectacle than the courage of a final moment, but it is no less a magnificent mixture of triumph and tragedy. A man does what he must—in spite of personal consequences, in spite of obstacles and dangers and pressures—and that is the basis of all human morality. . . .

In whatever arena of life one may meet the challenge of courage, whatever may be the sacrifices he faces if he follows his conscience—the loss of his friends, his fortune, his contentment, even the esteem of his fellow men—each man must decide for himself the course he will follow.

JOHN F. KENNEDY

Today we hear and read a lot about what professionalism is. In the fire service, a great deal of time is spent discussing what defines professionalism and how it can be achieved. Those who take the oath to become a firefighter take a pledge of service over self. For those who become fire officers, the badge they are given to lead and manage is a symbol that has become part of the fire service's tradition. This tradition is a wonderful heritage—bravery, loyalty, honor, ability, and devotion to duty resulting from a history of heroic deeds. The uniforms, the badges, and the insignias of the fire service symbolize this tradition. The badges and insignias are examples of a rich and historic symbolism, deeply rooted in a history of service. The cross displayed on many fire service symbols, known today as the "Maltese cross," was first worn by the Christian knights who shielded the weak. It is reflective of our modern fire service and has become a symbol of the protection of life itself. The badges that firefighters wear today are an outgrowth of the shields and coat of arms worn by the Christian knights to distinguish them as friend or foe in battle.

The emblem of the fire service is the cross of the Knights of St. John of Jerusalem, a charitable nonmilitary organization that existed during the eleventh and twelfth centuries. A white and silver cross on a dark background was adopted by these knights, or "hospitallers," as they were also known, because of their work in setting up hospices and hospitals for the sick and the poor. Later, they assisted the knights of the Crusades through their goodwill and by providing military assistance in an effort to win back the Holy Land from the Saracens. The Knights of St. John eventually moved to the island of Malta, for which the Maltese cross is now named. The emblems embossed on the shields of the knights became crucial markings in battle, as the armor they wore rendered them otherwise unrecognizable. The Maltese cross provided an excellent means of identification on the battlefield. Today, the symbols that

adorn fire service uniforms are a symbol of protection and a badge of honor, whose essence has been defined over the past 1000 years.

The knights, many of whom had a certain flair, often dressed in vivid colors to dramatize their presence. Many of them also wore large capes in addition to their suits of armor. These capes were made of crimson-colored cloth on the inside, reflective of the blood of Christ, and black on the outside, representing the knights' own sacrifice for humanity. These capes came to play a critical role on the battlefield.

The Knights of St. John encountered a new weapon, simple but horrible, unknown to European warriors at that time: fire. As they advanced on the walls of a city or fortress, glass bombs containing naphtha were tossed among them. Many of the crusaders were struck, becoming saturated with the volatile liquid. The breaking of these glass bombs signaled the Sacarans to hurl a flaming tree or some other burning object in their midst. As a result, hundreds of knights were burned alive. At the same time, the knights' capes became saturated with this liquid, so many had to become firefighters just to survive. Others tore off their capes and threw them over their burning comrades. During many battles, the capes of the knights of Malta saved many lives. In recognition, the Maltese cross became a medal of honor, a symbol of courage, and forever linked to the fire service. With the use of their capes, the knights risked their lives to save their brothers from dying fiery deaths. In essence, these men became the first in a long list of courageous firefighters. Their heroic efforts were recognized and rewarded by their fellow crusaders; each was presented with a badge of honor—a cross similar to the one that firefighters and chief fire officers wear today. The Maltese cross has come to represent the ideal of the fire service: the commitment to saving lives and property. Today this symbol represents all that firefighters do to protect their communities. It also represents their commitment to offering a good quality of life to the people who live and work within the jurisdictions they serve.

The trumpets that a chief fire officer wears on a brass collar or badge are symbolic as well. Their origin was first documented in 1752, when six speaking trumpets were purchased for the New York City volunteer fire department to enhance the ability of fire officers to communicate on the fire ground. The first trumpets were nothing more than megaphones constructed of tin, but soon other materials such as copper, steel, and brass were utilized. Each volunteer company enhanced its trumpet with insignias, silver plating, and markings to denote its identity. In the 1800s, chief fire officers utilized the trumpet to amplify their orders given on the fire ground. Today, the trumpet symbolizes the

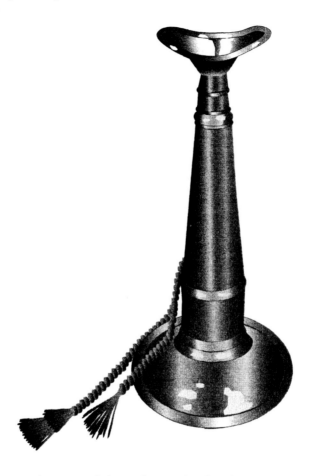

rank, authority, and responsibility of one who has been promoted in the fire service. Out of the necessity to communicate on the fire ground, a long tradition of fire service identification, recognition, and tradition began. The multiple crossed images of the speaking trumpet are forever linked to the ranks of chief fire officers. The trumpet is not only a symbolic remembrance of those who have gone before us, but is also symbolic of the roles and responsibilities that chief fire officers fulfill in carrying out their day-to-day responsibilities.

The North American fire service originated in local communities based on the tradition of neighbor protecting neighbor. With its roots in the colonial era, the image of the fire service recalls the bucket brigade passing pails of water from person to person toward the burning home. From these humble beginnings, the fire service has continued to evolve into a unique mix of highly trained men and women, both career and volunteer, who respond daily to a wide variety of hazards.

The evolution of the U.S. fire service is reminiscent of our history, from the colonial era to the Civil War, the Industrial Revolution, the post–World War I and II eras, and the technology revolution. The heritage and traditions

of this service have been instrumental in shaping the fire service of today. Many of the founding fathers of the United States were volunteer fire fighters, including George Washington, Thomas Jefferson, Benjamin Franklin, Samuel Adams, Paul Revere, Alexander Hamilton, John Hancock, John Jay, and Aaron Burr. In fact, Benjamin Franklin formed one of the first volunteer fire companies, the Union Fire Company, in 1736 in Philadelphia. A few years later, he set up America's first fire insurance company. He also organized a night watch and a militia to help keep the peace and ensure safety.

Today the public's perception of the fire service is molded less on fire service history and its' rich traditions or on the reality of what occurs from day to day, and more on what people see on television as the depiction of our industry. On programs such as *Rescue 9-1-1*, after someone dials 9-1-1, everything seems to go like clockwork. People respond quickly and arrive just in time to produce a positive outcome in every situation. Those perceptions clashed with harsh reality on September 11, 2001. Since that date, the perception of the fire service worldwide has changed dramatically.

For many reasons, the fire service will never be the same. The loss of so many firefighters in one incident, the courage they displayed, and their commitment to duty have changed the nation's perception of the fire service. The responsibilities and the role of leadership within the fire service have changed as well. There have been in the past, and will be in the future, defining moments in the fire service that help to create and frame what the industry is today and what it will become. Since September 11, I believe a new day has dawned for this profession, related not only to the challenges that the fire service faces but also to the responsibilities of those who will lead it. As never before, the fire service needs leaders who can articulate the needs of this industry at the local, state, federal, and international levels. Each day that passes is a day away from the tragedy that helped to redefine the fire service forever. And both those who are leaders today and those considering leadership roles in the

Source: Federal Emergency Management Agency.

future can never become complacent about what is necessary when they accept the badge of chief fire officer and the responsibility that comes with it.

I have witnessed many chief officers grow complacent and lose focus as they become mired in their day-to-day duties and operations. Yet, as chief fire officers, it is our responsibility to anticipate the future needs of our organizations. Equally important, we need to understand that we must not mistake how we do business today for what will happen in the future.

Don't mistake the edge of the rut for the horizon.

ANONYMOUS

This is an important statement. As we focus on our daily challenges, the rut is often an easy and comfortable place to be. However, in the future, the fire service will look very different than it does today, and those who have the ability to look beyond the horizon will be the ones who provide the architecture for the future.

Our industry cannot continue to approach its issues and problems with the same reasoning that has been used for the past two decades. As we look at our industry's mind set—the frame of reference from which we operate—we must begin to ask ourselves if we are limiting our ability to address the problems that most concern chief fire officers today. We must understand that by limiting our vision of what needs to happen, we also limit our ability to address critical problems and find the solutions to them.

As leaders, our frame of reference often dictates how we process information and thus determines what we believe is possible for the future. It is this belief that will determine how each of us will plan and move forward to implement that plan. Over the next several years, the fire service will need to create new frameworks and look at issues from new perspectives. Those of us who are leaders will need to remove the blinders that we so often use and have the courage to move our industry into uncharted waters. What the fire service will become in the next ten to fifteen years depends on today's chief fire executives and the fire service leaders of tomorrow. Not only will they be the architects of what is occurring today, but they will also be the bridge builders to what will be in the future. Today we have the opportunity to bridge the gap between the traditions that have made our profession what it is and what it will be in the future. That future will demand greater accountability and responsiveness, and a rethinking of how we approach our customers, the services we deliver, and how we deliver them. This will not be a simple task. It's often much easier to define the architecture than to actually construct it.

As fire service leaders, we must address many significant issues, yet we're often overcome by the complexity of the task. As a result, instead of taking incremental steps to achieve small successes, we have allowed the industry to avoid addressing the root causes of the problem. The significant questions for leaders in the fire service today are "What is our vision of the future?" "How will we get there?" "And who will provide the leadership to do it?" Many chief fire officers don't want to upset the status quo, even if they're locked into dysfunctional systems. Whether it's an organizational culture of bureaucracy or a culture that perpetuates poor behavior, the leaders of tomorrow will have to change their own attitude to produce a meaningful change in the fire service. The key issue is how we go about doing our business. It's really all about leadership. When a sufficient number of leaders decide to consider collectively our vision of the future, important things can happen. As leaders, we often know the direction that our industry must take, but orchestrating a major change is a momentous job. This is true whether the change occurs on a national level or in our own department. As we all know, it's impossible to do it alone.

I have often taken part in this process of change, but one experience stands out. In the mid-1980s, the International Association of Fire Chiefs (IAFC) formed a task force to develop a comprehensive evaluation process for the fire service worldwide. This was no small task. We were to create a process of voluntary assessment that could be used by any fire agency in any country, volunteer or career, in order to provide a system of continuous improvement. Two of the members, Ron Coleman and Ray Picard, were instrumental in its design and implementation. The program, as we know it today, is the Fire and Emergency Service Fire Accreditation Process. It is governed and administered by the Commission on Fire Accreditation International. This program has changed the face of the fire service. The number of spinoffs of this effort continues to multiply. The "standard of response coverage" concept has propelled the development of software programs to help fire agencies evaluate community risks. This has spurred the development of other software applications that link community risk factors to other geographical and informational databases, providing a more comprehensive analysis in designing response deployment for a given area.

CFAI Mission Statement

The mission of the Commission on Fire Accreditation International is to assist the fire and emergency service agencies throughout the world in achieving excellence through self-assessment and accreditation in order to provide continuous quality improvement and enhancement of service delivery to their communities.

Accreditation manager networks have been established throughout the United States to share information and compare analyses. Quality improvement has become an accepted methodology, and we now speak of things such as "performance measures," "benchmarks," and "best practices." These concepts have become part of our daily vocabulary. This process helps to drive standards on deployment and research at the federal level on firefighter safety, community risk, and the roles of the fire service today. This program prompted the development of a professional credentialing system for Chief Fire Officer Designation, increasing the professionalism of our industry.

CFOD Mission Statement

The mission of the Commission for Chief Fire Officer Designation is to assist in the professional development of the fire and emergency service personnel by providing guidance for career planning through participation in the Professional Designation Program.

Hundreds of fire agencies and chief fire officers worldwide are engaged in one or both of these programs. This effort evolved from a small group of individuals who went to the IAFC board of directors with the concept of building a different architecture for the future of the fire service. Individual fire chiefs have observed the same process in their own organizations. Individuals who join together with a common vision, purpose, and desire can make a great deal happen in a short period of time. Much of our future, like our past, will be dictated by those individuals who have vision and can make it a reality.

The fire service of the twenty-first century is ready to build on the impressive accomplishments of the past thirty years. Its abilities have grown exponentially in that time, as have the demands for its services. The fire service is simply no longer a local resource. Meeting the challenges of the future will take strong leadership and commitment from the chief fire officers of today and tomorrow. If we are to position the fire service to take full advantage of future opportunities, then we must be good stewards of the public resources that are given to us. This begins with our ability to understand where we have come from. The rich history and tradition of the fire service, coupled with the constant changes in the industry, will require leaders who understand that the challenges of our future will be met with innovation and tempered with the realities of our own experiences.

The Trumpets That We Wear

Trumpets adorn the uniform to distinguish the rank yet have a largely intangible meaning. They are emblematic of the qualities it takes to be a good chief fire officer. As you contemplate taking that step in your own careers, or if you are already a chief fire officer, you may find yourself struggling with the day-to-day duties that you're called upon to carry out. When I have days like these, I find it helpful to consider the badge and the crossed trumpets on our collars as reflecting the characteristics of successful chief fire executives—courage, integrity, a work ethic, and empathy. Throughout this book, I will touch on many of the characteristics that must be developed to be a successful leader. I will also explore the decisions and other related issues that a successful leader will have to address. If you are already a leader and wish to stay at the top of your game, what do you need to do to avoid becoming stagnant and ineffective? The unfortunate reality today is that many people who wear the badge of chief fire officer really aren't leaders. There is also reluctance among young fire officers to take the next step and become a chief fire officer for a variety of reasons. So, let's start by going back to the day of your graduation from the recruit academy, with your family and friends present to witness the pinning of your first badge. This is a proud moment not only for you as a graduate, but for your family as well. And as you look to promotion through the ranks and to possibly taking on the role of a chief fire officer, this has to be done with an acceptance of the responsibility, an understanding of the long tradition of service, and reflection on the day you received that first badge. The badge is a symbol of several hundred

Source: Courtesy of Linda Stone.

years of heritage and obligation. Unfortunately, many people have lost sight of this or simply don't understand it. The subtitle of this book, *A Symbol Is a Promise*, holds special meaning for me. I believe it is time that we consider what our badge represents, understanding the rich history and the traditions that have made the fire service what it is today. We must understand why our commitment to serve is based on a history that extends back almost 1000 years and why, more than ever, the symbols we wear are a promise to deliver.

Through the Eyes of a Nation

On the morning of April 19, 1995, an age of innocence was lost as a terrorist's bomb ripped through the Federal Building in Oklahoma City. As Americans focused on this horrible tragedy, the images they saw daily were of lives lost, of family members who would be seen no more, grief, and the struggle to survive, to understand why, and to reflect. Out of the chaos emerged a sense of order, calmness, and resolve to get through it.

The Oklahoma City Fire Department was thrust onto the world stage, and their performance was superb. Within hours of the incident, factual, professionally delivered media reports were given that kept all of us informed and made us feel part of the event. In the days and weeks that followed, Urban Search and Rescue (USAR) teams, and teams from other federal and state

Source: Federal Emergency Management Agency.

agencies, arrived from all parts of the country. The high level of professionalism displayed by all was seen in the images broadcast of the rescues made and of those that were not.

Technology has provided us with wondrous tools to look at places throughout the world and instantly become a part of life-changing events. With the flip of a television channel, the images of war, starvation, holocaust, tragedy, triumph, and victory are brought to us as they happen. With the ability to look at events in real time, the public's perception of the American fire service, fire services throughout the world, and the fire service's perspective on itself has changed and will continue to do so. Think back over the major incidents that have occurred within the past few years, when we in the fire service were critical of each other after watching events unfold on CNN and knew that things were not going according to plan. The question is "Does the public share that same perception?" Often it does not. Our fire companies have all been to a fire where they've created a parking lot by mistake. We've seen people die and thought that if only we had done certain things, the outcome would have been different. Yet, even when the building was burned to the ground, the owners or occupants still thanked us for our heroic efforts, even though we may not have done our best.

That's what impressed me about the Oklahoma City incident—it's where the world's perception met the reality of what we are called upon to do. As we watched this event unfold in the first hours after the explosion, and in the following days as we learned about the response of the Oklahoma City Fire Department and related agencies, it became apparent that the public's perception of their heroic efforts and the fire service's reality of a job well done were both well founded. In many situations we encounter within our own jurisdictions, this is not the case.

The images we are left with when a tragedy occurs can remain with us for a lifetime—the assassinations of John F. Kennedy and Martin Luther King, Jr., the explosion of the *Challenger* and the loss of the *Columbia* upon reentry. Most of us know exactly where we were when these events took place. The same is true of the images that emerged from the Oklahoma City incident—the firefighter leading a frightened and injured occupant down an aerial ladder saying, "Look at me, lean on me, trust in me, I will get you down." Those images, repeated time and again throughout this tragedy, will remain with the public for many years to come.

This was never more evident than on September 11, 2001, a day of tragedy for the American people and a day of great loss for the American fire service. It's also a day that reflects the subtitle of this book. The firefighters who responded in New York City, in Arlington, Virginia, and on the fields of Pennsylvania that

Source: Federal Emergency Management Association.

Source: Federal Emergency Management Agency.

day did so because they took an oath to serve. The unfolding of the flag on the side of the Pentagon was that same display from a national perspective. The flag, which symbolized our shared values, our freedoms, and our loyalty was a promise to the nation that, while bent, we would not be broken. The same holds true for those in the fire service and is reflected in the symbols that adorn our uniforms.

When firefighters take their oath, they are given a badge symbolizing the fire service. That badge is also a pledge that when a crisis occurs, they will respond and do everything they can to bring order out of chaos.

A fire chief has another core mission: to protect his or her personnel. Every year in October, the fire service comes together to honor those firefighters who

Source: Federal Emergency Management Agency.

have died in the line of duty. In October 2002 I attended the National Fallen Firefighters Memorial in Washington, D.C., where family, friends, and fire service personnel from throughout the United States gathered to recognize, remember, and honor all of the firefighters, both career and volunteer, who were lost in the line of duty the previous year. For 2001, due to the September 11 attacks, 442 were honored, 347 from the World Trade Center attack and 95 from other tragic events that occurred that year. The numbers don't tell the story, but the families do. Those firefighters lost, their ages ranging from the teens to the sixties, from different backgrounds, different parts of the country, and different departments, shared the same sense of pride, commitment, and dedication that make the common theme of service over self apparent to all of us in the fire service. Different as they were, they were connected in life by this bond that links everyone in the fire service. In death, they were connected by their families, friends, and colleagues who gathered to grieve, to heal, to share their experiences, and to remember, laugh, cry, and move forward.

The gathering of America's fire service to pay tribute to the men and women who have made the ultimate sacrifice is a time of healing and support. It gives them a chance to demonstrate to the families of those who died that we collectively share their pain and that they need not make this very difficult journey alone. It also refocuses our commitment on quality leadership and the understanding that for those who assume the chief fire officer's position, it is their time to lead. Part of that leadership involves making a commitment to improve the safety of firefighting personnel. As leaders of this industry, we must continue to take aggressive actions to stop the preventable accidents that lead to children without parents and families in pain. When we make a commitment to

Source: Courtesy of Linda Stone.

address this issue as leaders at the local and national levels, we provide a basis on which our profession will grow in the future.

Something significant happened on September 11 in the fire service. Not only did the world see, perhaps for the first time, the fire service's commitment and dedication to service over self, the fire service once again had the opportunity to see itself, and to understand just how deep the commitment to serve is and how important it is to our communities and to our nation. Since September 11, it has been interesting to observe the newfound respect for the fire service from elected offices and the general public. It's also interesting how we in the fire service have come to perceive ourselves. In many respects, we have reconnected to that thousand years of history. We have refocused on our mission—the protection of life and property and the core value of placing service over self. This reflects back on our true beginnings—the Christian knights and the Knights of St. John, who not only helped each other but also took up the call to serve others. That is what we do, and that is why it is so important that we understand today, more than ever, that the symbols we wear are truly a commitment and a promise to deliver to those we serve.

CHAPTER 2

Is a Chief Fire Officer's Job
the Right Job for You?

You must see your goals clearly and specifically before you can set out for them. Hold them in your mind until they become second nature.

Les Brown

I love football, especially college football. As someone who grew up in the State of Nebraska, it is almost a religion on a Saturday afternoon. For me, it represents what is good in life: teamwork, competition, focus, and the goal to be the best. There is a wonderful movie about college football that says a lot about those attributes, but also about personal commitment, and that is *Rudy*. *Rudy* is the true story of Daniel "Rudy" Ruettinger, who as a child made himself a promise that one day he'd play football for the University of Notre Dame. To do so he had to overcome tremendous obstacles, including his relatively small size, his lack of athletic ability, and the fact that he came from a blue-collar family with little money and minimal academic preparation. He never gave up on his dream, and he spent years working to achieve the grades and earn the money to go to Notre Dame. Once there, he joined the Notre Dame practice squad with no real hope of ever having a chance to play. Day in and day out, he suited up to take punishing hits from those with much greater size and athletic talent. Yet, day in and day out he came back, hoping for a chance to actually get into a game. If you've seen the movie, you know the impossible happened. Rudy got his once-in-a-lifetime opportunity to play during the final minute of a game. Lining up at a defensive end position, he blew past the blocker, sacking the opposing teams' quarterback for a loss. At the end of the game, Rudy was hoisted on the shoulders of his teammates and carried off the field in a triumphant tribute not to his prowess as a football player, but to his commitment, his persistence, his character, and his heart.

I think Rudy is a great example of what can happen when one makes a declaration. Rudy had the courage to do this and the perseverance to take a stand and focus on his commitment. Regardless of his circumstances, he made his declaration his reality. No matter how impossible a goal seems, the first step toward achieving it is the ability to declare the future. I have witnessed many declarations, for myself and for others, become reality. I know it works! When I became a fire chief at the age of thirty-four, I had made that declaration of a career objective many years before. When I became president of the IAFC, it was not by chance. It was a fulfillment of the declaration that I had made to myself and my family a decade earlier. By doing so, I allowed myself to prepare and position myself to succeed. Making the impossible possible is what declarations are all about. Today, many in the fire service often look at a position or a career track and say, "There are so few positions that I will never be able to compete." Success starts with a declaration: this is what I want to achieve. As we have seen historically, a declaration can become reality. When President John F. Kennedy announced on May 25, 1961, that "I believe that this nation should commit itself to achieving the goal, before this decade is out, of landing a man on the moon and returning him safely to the earth," many scoffed and said it couldn't be done. Yet, within a decade, Neil Armstrong set foot on the moon declaring, "One small step for man, one giant leap for mankind." When Robert Galvin, former chairman of Motorola in the 1980s, stated that Motorola was going to be the world's best company and defined a vision of six sigma quality (that is, fewer than three defects per million), many laughed and said it couldn't be done. But in less than a decade it had been achieved. Look at the pioneers of the computer industry—Tom Watson of IBM, Steve Jobs of Apple, Bill Gates of Microsoft—who all declared a new world through the possibilities of computerization. Within twenty-five years, computers and information technology have become an integral part of our lives. In fact, we cannot return to a precomputer existence. Each of these persons, in his own way, through commitment and vision, altered the course of history. In the same way, making a declaration is an important step if you want to compete for the role of fire chief and to make a difference for the fire service.

Do You Have the Right Stuff?

The movie *The Right Stuff*, which focused on the *Mercury* and *Gemini* astronauts of the 1960s and 1970s, promoted the popular image of these early astronauts. These men liked to push the envelope—driving fast, flying the highest, and living life to the hilt. All were test pilots who had immersed themselves in the technology of the day and were willing to sacrifice their own lives to be the first to challenge the outer atmosphere. It is through the *Mercury*, *Gemini*, and *Apollo* missions that Americans were able to witness what leadership and grace under pressure were all about. The confidence of the astronauts was due in part to knowing that they had a great team around them—flight directors, mission control experts, and technicians who watched every piece of data at mission control. They were led by mission commanders, whose job was to look out for the well-being of the crew, the ship, and the entire mission control team. One of those mission commanders, Gene Kranz, stood out. His reflections on his days at the National Aeronautics and Space Administration (NASA) involve a great deal of history as lived by those involved in mission control, and become the title of his book, *Failure Is Not an Option*. It was his mantra during the Apollo 13 flight crisis. Many from my generation still remember hearing the words of Captain James Lovell, commander of the ill-fated Apollo 13 mission: "Houston, we have a problem." We all know what has transpired since then, but the important message for those of us in the fire service is that leadership and teamwork are vital in overcoming adversity. For the *Apollo 13* mission command team, "Failure is not an option" became a shared value.

The story of NASA mission control teams as they guided the *Apollo* spacecraft through successful lunar landings and helped to save the lives of the *Apollo 13* crew is a never-to-be-forgotten chapter in the history of space flight but also offers an insight into the world of teamwork and leadership. It has been said that there are three kinds of people in the world: those who make things happen, those who watch things happen, and those who wonder what happened. Rudy Ruettinger, Gene Krantz, John Kennedy, Bill Gates, Tom Watson, and Steve Jobs are people who made things happen. They did so through visions and declarations, and by having the ability to create a shared value system with a core group to get things done. A chief fire officer must make things happen. What type of individual are you? Do you have what it takes to be a good leader? Do you have the right stuff?

Most fire chiefs today can tell the story of an intelligent, skilled fire officer who is promoted to the position of fire chief, only to fail. They can also tell the story of someone with solid but not extraordinary intellectual ability or technical skill who, on becoming fire chief, achieved great success. These stories support the belief that identifying individuals with the right stuff may be more of an art than a science. Max DePree observes that "Leadership is more tribal than scientific, more a weaving of relationships than amassing information." I have observed this to be true. The most successful leaders are those who can build lasting relationships rather than cite the management theories of different leadership styles. After all, personal styles of leaders can vary greatly. But there is a common thread among those who succeed. Successful leaders made their declaration early in their career, understood the enormity of the position, and made the commitment to prepare themselves, from both a relational and an educational perspective, to succeed. Part of having the right stuff is knowing the right questions to ask and the right direction in which to move. The most important day in your life can be the day you start moving in a new direction: the day you learned how to ride a bicycle, received your driver's license, graduated from high school, got married, witnessed the birth of your first child, became a firefighter, and the day you were promoted to lieutenant or captain. These are significant life events that forever changed the direction of your life. It is important to have as many positive significant life events as possible. The role of fire chief, or another senior command position, may then be the right one for you.

Let's consider the job of senior command officer. One of the things that is very difficult for many new chief officers is the separation between them and their former colleagues, whom they must now lead and manage. And in many organizations, the person who becomes a chief officer is no longer part of the rank and file. In fact, on both sides of the equation, we tend to create separation between labor and management, ranging from social activities to the normal activities that occur around the fire station. One of the most difficult adjustments that many new chief officers have to make is to understand that a difference in relationships will emerge due to the natural order of things.

I know this was true for me. It took me more than a year to understand the new dynamics in the relationships, and it probably took me just as long to adapt. In the course of my career as a fire chief, I have observed many of my battalion and deputy chiefs struggle with this transition. They try to keep one foot in both camps: as leaders and as one of the guys. This does not work. Senior managers who shift back and forth on issues lose command authority and lose the respect of their subordinates, their peer group, and the chief.

Another factor that is important to understand is the demands that being a fire chief place on your family life. Many of us have come up through the ranks. We have worked 24/48-hour schedules or an alternative schedule if assigned to a staff position. In either case, when we left the firehouse we were able to leave the job behind. This is not the case for a chief fire executive. Being a fire chief is a 24/7, 365-day-a-year commitment. For many, that can be an overwhelming burden and one that can cause a great deal of stress at home. That is why the decision to move up within the fire service and to tackle the chief fire officer position is a decision to be weighed not only by you but also by your spouse/significant other and family as well. It may even influence where your family lives, and will create a new set of stresses and strains for everyone.

But for many of us, the desire to be a leader is a calling. Regardless of the perceived downsides of being the chief fire officer, people are still drawn to the leadership role because they know *they can make a difference* or they just want to be in charge. For those who fall into that category, motivation is not the issue. Preparation and our abilities are more often what they struggle with. One of my battalion chiefs pointed out that he didn't think "we were growing as many fire chiefs as we used to." He had observed that in the younger workforce today there do not seem to be as many Type A persons as in the past. Further, those who are not Type A don't necessarily need to be in charge to gain satisfaction from their job. He observed that many want a better balance between work and the rest of life. They just want to go to work, do a good job, go home, and not worry about it. Those of you who fall into that category should read on; the fire service needs you!

Fire chiefs have the ability to impact a community's fire protection. Through their position, their commitment to fire protection, and their personnel, they can make a difference in the lives of the citizens they have been chosen to represent, and they can impact positively the people they have been hired to lead. Looking back over my career and the number of people I have promoted and demoted, hired and fired, the number of fire protection ordinances that have been passed while I've been fire chief, the innovations and equipment and facilities that I have helped to acquire, it's clear that I have had a significant impact on a number of people. I am just 1 of approximately 30,000 fire chiefs in the United States, and there are thousands more throughout the world. In each case, those who assume this position must do so with a dedication and commitment to make a difference. For me, it goes back to a couple of days before I was to graduate from the firefighter recruit academy I was attending in Aurora, Colorado. I was called in by Assistant Chief Dick Meisinger and Lieutenant Carl Smith, who were running the academy. When this happens, it is usually to inform the person about a lack in some area of development, but that was not the case here. They had called me in to tell me that they saw more in me than, at that time, I saw in myself. They told me that at some point in the near future I would be a fire chief if that was my desire, and encouraged me to begin to prepare myself to take on that role. When I left the room, my mind was spinning, spinning concerning the possibilities. That was wonderful mentorship and a lesson for all of us in leadership positions.

I've found myself many times in that same position on the other side of the table, encouraging young firefighters and officers who have the talent to be chief fire officers to begin preparing themselves for that role. I have seen many people wait until a couple of months before a promotional test or a fire chief recruitment to begin to prepare. I often wonder: if Chief Meisinger and Lieutenant Smith had not alerted me to the possibility, would I have had the sense and the insight to do it on my own? I don't know. But I do know this: no matter how impossible you think your goal is, if you don't imagine it and don't declare it, you will never develop a game plan to achieve it, and you'll never arrive at it. This is the story of Daniel Rudy Ruettinger. His dream of playing football for Notre Dame came true because of his declaration to do it, his commitment to make it happen; that gave him the persistence to achieve it. It's no different for others. Once one has made the commitment to make a difference, to set a course, to achieve a goal, great things can happen.

What Does It Take to Be a Good Leader?

The complexities that chief officers face today in leading and managing fire service delivery systems are defining a new set of characteristics and leadership styles for the industry. To be a good chief executive is a tough and demanding job. Many aspire to the position without having a clear understanding of the commitment it will take, or the demands of leadership, until they have actually accepted the responsibility. Often this creates an unfortunate situation for the agency they have been appointed to lead. As you contemplate moving up to the next level, or if you are already a chief fire officer, you must ask yourself what qualities make for a good leader. You probably know, have worked for, or have been introduced to people you thought had good leadership qualities. What were the qualities that made you take notice of those individuals? We are always attempting to define good leadership. In a 1998 *Harvard Business Review* article, Daniel Goleman observed that leaders need vision, energy, authority, and strategic direction. I would agree, as I have never met an effective leader who did not possess those attributes. He also stated that successful leaders share five important qualities that separate them from others and provide them with the tools to maximize their leadership potential:

- They selectively show their weakness by exposing vulnerability: self-awarness.
- They rely heavily on intuition to gauge the appropriate timing and course of their actions: self-regulation.
- Their ability to collect and interpret soft data often helps them to know when and how to act, or react, to a situation: motivation.
- They manage employees with something they call "tough empathy." Inspirational leaders empathize passionately and realistically with the employees they care intensely about and the work they do: empathy.
- They reveal their differences, and are able to capitalize on what is unique about themselves as leaders: social skills.

These qualities separate those in the fire service who are leaders from those who hold a position of leadership. They set great leaders apart from the rest. I know this to be true, as I have had the opportunity to work with the best fire service leaders, and the characteristics noted by Goleman are evident in each of them. I also think there are several other traits that are paramount for those who lead in the fire service if one is to be successful and have the impact that leaders should have. These will now be discussed.

The Courage to Lead

Courage is the first of human qualities, because it is the quality which guarantees all others.

WINSTON S. CHURCHILL

It's difficult to lead an organization today, whether a fire department or a small business down the street. The chief fire officer often feels under a microscope on a daily basis. In comparison to the small business owner, the fire chief's ability to produce outcomes, whether measured by profit or lives and property saved, is measured daily and weekly. For both, talent and ability is what ensure success. In the fire business, we see courage on a weekly basis in the lifesaving actions of our firefighting personnel. But I have also witnessed many heroic and courageous efforts by chief fire officers. One fire chief, after a mayoral election, was reassigned to another department so that the mayor could repay a political debt with the appointment of a new fire chief. That ex-fire chief keeps serving, with honor, the residents and the mayor of the community in another position. That's courage. Another chief fire officer proposed a sprinkler ordinance, only to find himself in the crossfire of confrontation between the builders and elected officials, at one point having his job threatened by the mayor if he did not back down. It was his courage and convictions that persevered to gain a comprehensive sprinkler ordinance that will save lives and reduce firefighter risk for many years to come. That's courage. Still another chief fire officer, in conjunction with the city management team, had negotiated an agreement with the labor group, only to be told later that day by the city manager that he had changed his mind. The city manager told the chief to inform the union president that the city manager had changed his mind. The chief resigned the next day, citing ethical reasons. That's courage. Remember, you may wear the badge, but without courage you won't be the fire chief.

Integrity

Integrity in all things precedes all else. The open demonstration of integrity is essential.

Max DePree

The leader's self control sets the example for followers. In discussions of leadership, it has often been said that if you can't talk the talk and walk the walk, then others will not follow. In the field of public safety, integrity in all that we do is an absolute necessity. What do I mean? Let me give you an example. Should the fire chief who sends a group of personnel to a conference expect them to spend a day playing golf at the organization's expense in the community where the seminar is held? Probably not. Yet many fire chiefs do so. Are we sending the right message to those we are charged to lead? After a while, many can rationalize crossing the line to justify their own actions. The reality is that the communities and the people we serve often have a higher expectation of public servants than they do of themselves in their own business settings. For the public safety official—fire, police, and emergency medical service (EMS) personnel—the expectations are even higher, as they should be. After all, we are the ones they call upon when they are in crisis. We enter their homes, are given personal information, and are privy to the most sensitive information. Sometimes this makes it difficult for us to meet those expectations. For all of us in the fire service, those expectations are a reflection of the trust that our profession demands. When we fail to meet them, the incident often becomes the lead story on the local news. Unfortunately, when that occurs, everyone in the fire service is harmed. The perception of misconduct or improper activity affects everyone in government. We have seen this occur at the local, state, and federal levels.

While I lived in Oregon, the state was experiencing significant shortfalls and unfunded liability for the public employee retirement system. The cause of the problem was multi-faceted—a stock market crash, poor administrative oversight, and unwillingness by the appointed board to address the tough fiduciary issues. Also, at that time the pension board was composed predominantly of public employees. The point is how this situation affected the attitude of the general public. Headlines such as "Government Workers Line Up to the Trough" and numerous editorial cartoons left the impression that everyone in government was either unethical or overpaid and was taking advantage of the system. This general perception has led to continued mistrust of state government and has been reflected at the polls by the failure to reelect incumbents and by the defeat of taxing proposals to keep schools open. And in the federal government there are many examples of national misrepresentations such as Watergate, Iran Contra, and the Clinton impeachment process, just to name a few. These all fuel the skepticism surrounding our government.

The mistrust of government today, and of the people in government, is high, and the root cause is often found in stories such as the ones above. It also goes beyond what is visible to others. The actions of chief fire officers in dealing with discipline issues, promotions, and elected officials (the list is endless) will speak volumes to their integrity. In the messages they send and the actions they take, integrity is essential.

The Willingness to Assume Responsibility

As chief fire officers, despite either good or bad news, we have to be willing to assume responsibility for what has occurred in the organization and for those under our command. Though this is often difficult, as events are sometimes beyond our direct control, that is what leading an organization is about—responsibility. The reality is that it all starts with the expectations the chief sets as a leader. When faced with these situations, the chief's actions will say much to employees and to the community about how he leads and manages the organization.

Being fair and just is a critical element of integrity. Leadership requires a keen sense of what is right and what is wrong. Whether the issue is ethical, moral, or disciplinary, the ability to display fairness in making decisions is extremely important. There are many things in our industry that are black and white. From a leadership perspective, in those cases, it's much easier to be fair and just and to provide leadership. There is often a clear road map to follow. But chief fire officers often deal in shades of gray. Fairness lies largely in the ability to articulate your position on why you have chosen to take a certain path. Leadership is also the ability to get to the heart of the matter, and once there, to have the integrity to do the right thing.

One of the most difficult things for leaders is disciplining a subordinate who errs. An example of accountability, yet poor leadership, was a report I was given regarding a minor vehicle accident a couple of years ago. The accident resulted when the apparatus operator turned too sharply and ran into an object, causing minor damage to the vehicle. It was clear from his report, which indicated that he had cut the corner too close, that it was an avoidable accident, and that he was willing to assume the responsibility and learn from the experience. Conversely, his officer had stated in his own report that major damage was avoided because of the apparatus operator's quick reaction. It is interesting how different perspectives can be brought to bear on the same incident. It also provides a good example on what makes for poor leadership. In this case, the company officer was doing nothing more than trying to cover for the apparatus operator. That's not leadership. It is nothing more than abdicating the responsibility of leadership. This is an excellent example of reconfiguring the facts to reach a plausible and acceptable conclusion to avoid having to make a tough choice and deal with the issue.

To be a successful leader, two key characteristics are fairness and the willingness to tell the truth, whether it makes people unhappy or not. One of the most difficult issues that leaders face is dealing with personnel problems. To be honest, many in the fire service are not effective in this situation. This is due largely to the types of shifts that we work and the relationships that develop when people live together for a third of their lives. It becomes much more difficult to discipline someone who is often your best buddy. Many people who move up through the ranks can never let go of their relationships and be leaders. The end result, as evidenced in the previous story, is a person in a leadership position who is ineffective, not well respected, and easily manipulated by subordinates based upon relationship factors. For these people it also creates significant frustration, as they are continually in conflict with themselves; they know what they should be doing as leaders but cannot bring themselves to do it.

At the other extreme are the people who, once promoted, in order to prove themselves leave a path of destruction in using discipline. These are the people who approach the job thinking that if they display their authority, they will be seen as a good leader. I have worked with a number of them as well, usually as the "relief pitcher." These people, due to lack of interpersonal skills, because they are trying hard to separate themselves from their former lives or trying to "mark their territory," are so authoritative, and in many cases so myopic, that they lose sight of the real issue.

Work Ethic

Nothing in the world can take the place of persistence. Talent will not; nothing is more common than unsuccessful men with talent. Genius will not; unrewarded genius is almost a proverb. Education will not; the world is full of educated derelicts. Persistence and determination are omnipotent.

CALVIN COOLIDGE

Every successful leader I have met has a strong work ethic and a willingness to do more than he or she is paid to do. In fact, money is never the issue with these people. If you wish to become a chief fire officer, you have made the commitment to do more than you demand of the people you lead. If you're not willing to do that, how can you ever hope to establish a "can do" corporate culture in your own department? The leader of an organization who expects professionalism, high performance, and productivity is probably a leader who is willing to sacrifice for the organization and the people in it. I have worked with others who had an entitlement mentality. Their purpose was to see how much they could get out of the system. There are many such people: the chiefs who never produce or are gone all the time or the twenty-year veteran who spends more time on injury or sick leave than at work and has a continual claim at the worker's compensation board. Unfortunately, it is a trap that some fall into. I repeatedly ask myself, "Am I adding more value than I cost the organization? Does my insight and my ability to negotiate and problem-solve save the department money? Because of my leadership abilities, is the organization exposed to less risk? Because of my leadership, are people's talents aligned with the needs of the organization?" I think the answer to each of these questions has to be yes, or you should step back and look at your own career track.

Today we often hear that the fire chief is routinely on a three-and-a-half-to four-day week and that the productivity requirements for the chief fire officer and his or her senior staff are much less than those of others in the organization. If you find yourself falling into this category as a chief officer, or if you're already in that category, you need to rethink your mode of operation. It puts you in a very vulnerable position, one that is not a characteristic of a true leader. Just as importantly, you have to ask yourself why you are in that position and what has led you to become so complacent. If you are a chief in that position and you are comfortable with it, it is time for you to do something different. This is a tough decision, but your value to the organization is gone and in fact you are probably doing more harm than good. The other reality is that you are probably not happy either; the change may be good for both you and the organization.

Empathy

Some people think only intellect counts: knowing how to solve problems, knowing how to get by, knowing how to identify an advantage and seize it. But the functions of intellect are insufficient without courage, love, friendship, compassion and empathy.

DEAN KOONTZ

Chief fire officers need to be attuned to the needs of the people they lead, but they must also understand the complexities and the diversity of the organization. It's easy to come up with an idea, throw it on the table, and expect people to achieve immediate results. As chief fire officers, after all, that's what we're paid to do. But as the leader of the organization, we must understand the ripple effects of our decisions throughout the organization. Whether you're the fire chief or a battalion chief, never underestimate your influence on the people who work with you. An idea you propose or a direction you indicate can set a dynamic in motion that has shiftwide and/or organizational impact. Good leaders also know the synergy that can be created on the back step of a fire engine or at the kitchen table in regard to the relationships developed within the organization and with the leader.

Organizations are much like a balloon full of water. If you squeeze it on one side, it distends on the other. If you overfill it, it will burst. If you underfill it, it's much easier to pop. Good leaders know when to add water and when to take some out, when to push and when to cradle. That begins with having a good understanding of what is happening within the organization. The same holds true for the people being led. The leader has to know them well enough to know when to push and when to pull, when it's time to be compassionate and when it's time to be tough.

Communicator

As a chief fire officer, you will be challenged to bring a global perspective and a vision to the organization, yet at the same time to be a master of the details. Your credibility will often depend on taking the time to edit a document before it goes to your board, council, or to fire chief. You've heard it said before: "the devil is in the details," and it is. In many respects, the chief fire officer's role is to make sure that the details have been articulated, researched, and effectively presented so that the governing body can make a quality judgment and a quality decision for the organization.

Many times, not paying attention to details that were the responsibility of someone else has placed many fire chiefs, including myself, in a precarious situation in which they were trying to explain poorly documented facts and analysis to an elected official or their superior officer. On numerous occasions, I have had to help bail out a senior officer at a meeting of elected officials because the report presented was so poorly done that it brought into question the credibility of the report and the presenter. Once that occurs, it is like a feeding frenzy at the zoo—everyone wants a piece of the action, and the line of questioning can often get very uncomfortable.

The development of a cooperative atmosphere with others is a positive outcome of quality communications. It is essential to the chief fire officer's success. Whether internal to the organization or external, the chief's job is often about creating partnerships and relationships to get things done. Leaders are responsible for being effective, doing the right things and doing things right for their people. Great leaders understand that people are the driving force within the organization. Our challenge is to get all the parts of the machine moving and producing (outcomes) in the same direction. Good leadership encourages contrary opinions, allows people to have their own space, a sense of freedom that enables them to grow. This encourages them to open up to new ideas, to include the diversity of opinions within our organizations. Often this is not an easy task, but nonetheless it is extremely important.

The chief fire officer can't bring a global perspective and a vision to the organization, or create cooperative relationships, without being a good communicator. Effective communication is the key to being focused and providing energy to implement a vision. Remember, you can have the best idea, but if it's not communicated throughout the organization in a way that people can accept and support, it's worthless.

Communication also involves understanding the perceptual biases that we all have as we communicate and process information. As you read through the list on page 33, can you think of situations you have experienced that were defined by one or more of these perceptual biases?

Here are some basic strategies to avoid miscommunicating your vision and strategy:

- Use different forms to spread the word. You can't send out one memo or one videotape, have one conference call, or go to one meeting and expect people to begin to understand the strategy and vision that you're trying to communicate. You have to use multiple media. If people are hearing it, seeing it, and beginning to live it organizationally, then they can grasp the vision and help to support it.
- Repetition is the key. As with any good marketing strategy, the more you hear it and see it, the more likely you are to be able to repeat it. Therefore, repeat, repeat, and repeat the message that you are trying to communicate within the organization.

Perceptual Biases	
1. First Impressions	We form them almost instantly.
2. Stereotypes	We tend to overgeneralize and categorize.
3. Selective Perception	We see and hear what we expect. We seek information consistent with our own views.
4. Concrete Information	We weigh more heavily things that we experience directly or that are more vivid to us.
5. Negative Information	We weigh and accept negative information more easily than positive information.
6. Inconsistency	We believe we are more consistent than we actually are.
7. Law of Small Numbers	Small samples are assumed valid. A few cases "prove the rule."
8. Complexity	We tend to simplify and have trouble handling complex information, especially when we are under pressure or time constraints.
9. Gambler's Fallacy	We tend not to understand the laws of random chance. We believe chances for rare events improve over time, when they don't actually change at all.
10. Availability	Things we easily recall are overestimated.

Source: Adapted from McCall and Kaplan (2001).

- Lead by example. If you do the opposite of what you say, no one will listen to you.
- Keep it simple. If you're communicating a new vision within your organization or a new direction, present it in such a way that people can grasp it. Many vision statements are so complex that ten different employees interpret them in ten different ways. Simplicity is a must if people are to understand the vision and get behind it.
- Last, and maybe most important, is listening. To be effective, communication has to be two-way. We often get so caught up in communicating our vision, message, or new idea that we fail to listen to the feedback that is given to us. That feedback can be critical, especially if our workforce is asking, "What does this mean?" When Ford Motor Company implemented its quality improvement process, management built their campaign around the theme "Quality Is Job #1." It became a slogan, a

means to communicate, not only to their workforce but also to the world. They stressed to all that Ford had embraced a new vision to improve the quality of their products, and it worked. One of the keys to success was that they were able to communicate to everyone in the organization what it meant to each of them. Conversely, the employees were able to communicate back in regard to how they saw their role and responsibility in relation to the vision of "Quality Is Job #1." What was created was alignment of thought and action brought about by good communication.

Profiles of Leadership

Being a good leader today means more than being in touch with your own reality; it's also about understanding the reality of others. Good leaders have the ability to understand what it is like to walk in the other person's shoes. So, when changes have to be made or new directions set, they have a good sense of the impacts and the behavioral response that are likely to occur.

Father Miles O'Brien Riley of the Archdiocese of San Francisco offers an insightful view of what it takes to be a quality leader from his series on values, ethics, and leadership.

Is It Time to Make a Declaration?

Father O'Brien offers a different perspective on what it takes to be a good leader today. His profile of creative leadership emphasizes, courage, integrity, responsibility, a strong work ethic, empathy, and the ability to communicate. Each is a critical element for those striving to be an effective leader.

As I previously stated, once you realize that you want to be a leader, you must begin today to prepare yourself to take on that role. The first and most important step is to sit down with your spouse/significant other and your family. Make the declaration that you want to become a fire chief officer and, to do so, outline the steps that might be involved. It may mean that your family has to move, or that you must accept a little more stress than you have today, but it offers a great opportunity for you to impact a community and to influence positively the people you lead. It starts by asking yourself if you think you have the right stuff—to lead, to take responsibility for the community's fire protection, to lead organizational change, and to manage and balance the stress that will occur as a result, not only for yourself but for your family as well. You must ask yourself if you can find a balance so that you maintain a happy family life, and a degree of sanity, and enjoy the journey. Once you declare the future and take a stand, your world begins to transform. You come to realize that anything is possible, no matter how big your goal is or how impossible it seemed just a few months earlier. When you can declare your future, you can make it happen. If the future for you is to become a chief fire officer, begin to prepare today. Make the declaration, commit to it, and have the heart to make it happen. We will need good leaders in the future.

Profile of a Creative Leader	
Leadership (Self-less)	The creative leader is dedicated to *making someone else look good* without making himself appear primarily responsible.
Harmony (Integrity)	The creative leader has *good emotional control:* vindictiveness, irritation, sulking, rejection, defensiveness, despondency and emotional outbursts are uncreative.
Free (Change-growth)	The creative leader is *independent:* not aloof or unconnected—but self-starting, self-motivated, and flexible.
Self-Control (Restraint)	The creative leader is *dominant: not controlling* others but mastering himself, not manipulative but self-directed.
Self-Confident (Humble)	The creative leader is *confident:* he knows and accepts himself *with humility*, and without apology or anxiety: he strives to improve himself, his skills, his talent: he had the *confidence to operate behind the scenes.*
Honest (Truth-talk)	The creative leader is *truthful:* he describes reality with flair and verbal artistry but as it is.
Candid (Approachable)	The creative leader is *open:* he speaks with spontaneity, enthusiasm, clarity and friendliness; he is open to new ideas, suggestions, and advice.

(Continued)

Profile of a Creative Leader (Continued)	
Energetic (Responsible)	The creative leader is *energetic:* he is a professional *learner*, absorbs vast amounts of information, has a need to be productive—he is *response-able*.
Communicates	The creative leader is a *communicator*: he has a respect for the power of words, he cares about accuracy, meaning and motivating others.
Executes	The creative leader is an *executive:* he *executes* not as a scribe but as a creator of positive images, he fights for the truth, he advises fearlessly without self-interest.
Loyal Compassionate Happy	"The creative way of genius demands sacrifice." (N. Berdyaev)
	Courtesy of Father Miles O'Brien Riley, Ph.D. Director of Communications Archdiocese of San Francisco

CHAPTER 3

Developing Your Game Plan
for Success

Success is peace of mind, which is a direct result of self satisfaction in knowing you did your best to become the best you are capable of becoming.

JOHN R. WOODEN

You've made the decision to become a chief fire officer. Probably the best way to start is to read all the trade journals and become familiar with the requirements to compete for the position. In the past decade, I have seen a subtle shift in the requirements to even be considered for the position of fire chief. Many of them are the tangible criteria of education and certifications. Others are more subtle, concerning the experience and background that city managers and mayors are looking for, as well as the political processes used within organizations. Let's talk about education. It used to be that if you had either an associate's or a bachelor's degree, had a number of years of experience, and held the rank of captain or above, you could compete. With the right presentation, you could probably get an interview for a fire chief's position. But that is not the case today. Over the past ten years, a significant shift in the requirements has occurred. Most positions require at least a bachelor's degree, and a master's degree is preferred. Many city managers as well as recruiting firms are now looking for Executive Fire Officer certification as well as the Chief Fire Officer Designation. The educational bar has definitely been raised, so if you are considering or have decided to become a fire chief, then you need to begin to develop your game plan accordingly.

Today a master's or Ph.D. is something you should be considering at some point in your career. Our officer corps is much better educated today and will be more so in the future. If you already have an advanced degree, you should expand your certifications and become involved in three specific programs: the Executive Fire Officer Program (EFOP), the Commission on Chief Fire Officer Designation, and the Institute of Fire Engineers (IFE).

The Executive Fire Officers Program

The EFOP is an initiative of the United States Fire Administration/National Fire Academy designed to provide senior officers and others in key leadership roles with:

- An understanding of:
  - the need to transform fire and emergency services organizations from reactive to proactive, with an emphasis on leadership development, prevention, and risk reduction;
  - the need to transform fire and emergency services organizations to reflect the diversity of America's communities;

Source: Courtesy of FEMA/United States Fire Administration.

- – the value of research and its application to the profession; and
- – the value of lifelong learning.
- Enhanced executive-level knowledge, skills, and abilities necessary to lead these transformations, conduct research, and engage in lifelong learning.

Officers enhance their professional development through a unique series of four graduate and upper-division baccalaureate-equivalent courses. The EFOP takes four years and has four core courses. Each course is two weeks in length. EFOP participants must complete an Applied Research Project (ARP) related to their organization within six months after the completion of each of the four courses. *Note: Completion of the ARP is a prerequisite for attending the next course in the sequence of the program.* A certificate of completion for the entire EFOP is awarded only after the successful completion of the final research project.

Chief Fire Officer

There are many fire chiefs and persons who use similar titles, but only a select few can qualify to be designated a chief fire officer (CFO). These individuals have demonstrated through their education, leadership, and management

Source: Courtesy of Commission on Fire Accreditation International.

skills that they possess the requisite knowledge, skills, and abilities required for the fire and emergency services profession.

Achieving professional designation as a CFO acknowledges that the individual has attained the status recognized by his or her peers, superiors, and subordinates. Other similar organizations within commercial, industrial, and governmental agencies also recognize this mark of distinction.

A fire chief, subordinate chief officer, senior emergency services officer, or individual assigned to a senior-level position may receive the professional designation of CFO if he or she successfully completes the requirements of the Chief Fire Officer Designation program.

The Chief Fire Officer Designation program is a voluntary program designed to recognize individuals who can show excellence in seven measured components:

- Experience
- Education
- Professional development
- Professional contributions
- Association membership

- Community involvement
- Technical competency

These seven components, as shown in the accompanying figure, encompass the professional and personal characteristic profile desired for a chief fire officer.

Designation Profile

The program consists of seven (7) components, as shown below, which encompass the professional and personal characteristic profile desired of a chief fire officer.

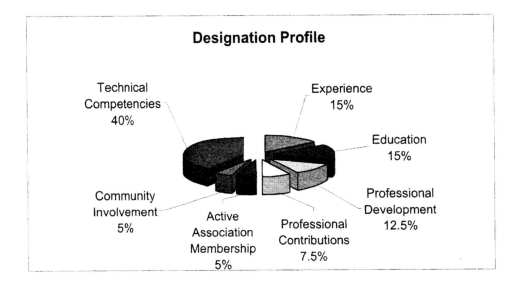

Component	Maximum Points*	Percentage
Experience	150 pts	15%
Education	150 pts	15%
Professional Development	125 pts	12.5%
Professional Contributions	75 pts	7.5%
Active Association Membership	50 pts	5%
Community Involvement	50 pts	5%
Technical Competencies	400 pts	40%

Total: 1000 pts

*The listed point values and respective percentages are the maximum for each component. While it is possible to accumulate more points based on application, each component ceiling represents a maximum award. Points MUST be accumulated in each of the seven components. 800 points is the minimum threshold required for designation.

Source: Courtesy of Commission on Fire Accreditation International.

The Institute of Fire Engineers

The Institute of Fire Engineers was founded in 1918 in the United Kingdom by a group of eight British chief fire officers. The mission statement they adopted continues to guide the Institute in meeting the needs of its members and serving the interests of society:

> *To promote, encourage, and improve the science and practice of Fire Engineering, Fire Prevention, and Fire Extinction, and all operations and expedients connected therewith, and to give impetus to ideas likely to be useful in connection with or in relation to such science and practice to the members of the Institution and to the community at-large.*

The United States of America (USA) Branch continues the tradition of fire service leadership in promoting fire engineering established by the forward-looking men and women who established the Institute. Another small group of fire service leaders met in Tulsa, Oklahoma, in February 1996 to establish the branch. Membership in the Institute of Fire Engineers is normally conferred through an examination process. Individuals who achieve good scores on the membership examinations are admitted to the appropriate grade based upon the nature and content of the examinations passed. The Institute currently administers the following examinations through its branches:

- Preliminary Certificate
- Intermediate
- Graduate
- Member

The USA Branch is not currently administering the membership examinations of the Institute of Fire Engineers in America. As such, the branch can only nominate members to the grades of Student, Graduate, Associate, Affiliate (Organization), and Fellow at this time. The USA Branch Membership Committee reviews each application and provides the necessary endorsements for those who meet the experience and educational criteria.

As you continue to expand your academic resume, try to include specialized classes in quality improvement, planning, or finance, all of which are vital for today's chief fire executive. It is extremely important to begin today to develop your game plan for the next ten years. One of the things I have observed is that many of my officers, even those who have the potential to move up, assume more responsibility, and expand their knowledge base, procrastinate. They delay their decision to either go back to school or take a class to obtain a new certification until just before the test. If you want to become a chief fire officer, you have to begin working seven to ten years earlier to prepare yourself. If you have not prepared educationally, experientially, and intellectually, you will not get the opportunity to be considered for the initial cuts, to get the interviews, to gain the experience, and ultimately to get the position. That's the reality of the game. So, if you have made the declaration that you want to become a chief fire officer, you need to map out a strategy. Begin by looking at the educational requirements today and how they match your own background and experience. Determine where your deficiencies are so that you can develop a course of action to begin solidifying your academic background, as well as your experience, in order to be competitive.

Let's discuss experience and the shift that has occurred over the past ten years. We have seen a shift away from a primary focus on operational experience to individuals with a more global perspective. Fifteen to twenty years ago, the fire chief was usually selected based upon the ability to fight fires. In many cases, the highest-ranking assistant chief or deputy chief in the organization who was in charge of operations was the automatic heir apparent to the fire chief's position when the fire chief left. That is generally no longer the case. Today, many city managers, mayors, boards of directors, and councils are looking for leaders who have a more rounded background and perspective. The changing landscape of government today requires a new skill set for those who will lead in the future. A broad base of experience in operations, training, administrative services, finance, and work in the fire prevention division, are now becoming basic criteria for recruitment. An individual who has spent time not only working a shift, but also working a forty-hour week in multiple positions, will have a more global perspective on how government works. Today, recruiters and government leaders are looking for persons who can bring to their organization a broad perspective, rather than a perspective and experience in one area of expertise. Experience outside the fire service in other government sectors and private business is also an advantage. Many candidates attempt to compete for a fire chief's position after having worked in only one division—operations, fire prevention, or training. Many have found that this impedes their ability to compete effectively against others with experience in each of these areas and many others. The chief fire officer of the future must possess global experience, as well as being well rounded and versed in all aspects of emergency services, as well as general government. Therefore, it is imperative to explore ways to gain experience in each of the areas of service—fire prevention, training, operations, finance, human resources, and so on. All are critical to the fire chief position. The person with experience in each of these areas will be much more marketable than one with a background in only one of them.

Find a Competitive Advantage

Now that we've talked about academic degrees and gaining more global experience, let's consider another important factor: finding a competitive advantage. Gaining a competitive advantage in the field of fire service has to do with finding a niche and gaining recognition and experience from it either locally, regionally, or nationally. Find an area that you can write about, teach, and articulate so that you can become recognized as an authority. I've watched people compete for fire chief positions who either had or lacked that competitive advantage. More often than not, those with an advantage were chosen for the position. Persons who can describe a program or a process to others so that they understand it and adopt or implement it are demonstrating leadership. This is very beneficial when competing for the position of fire chief. So, find a niche—something that you enjoy doing and do well, something you can communicate to others within the fire service or general government, something that you can write about and teach. This will differentiate you from the other candidates you will compete with.

Let's face it, in many cases we are all very similar. So, when we move through the competitive process, we often look similar in the way we approach problems. This is because we've been trained in very similar ways—in the way we present ourselves, in many of our academic credentials, and in related certifications. So, if you can gain a competitive advantage that makes you stand out, you are in a much better position to obtain the job you desire. To do this, it's extremely important to make a declaration to be a fire chief and begin immediately to develop a game plan. That includes gaining academic degrees and certifications, and formulating a plan to obtain that global experience we talked about. It's also about finding your niche and developing a competitive advantage over the next several years so that when you begin to compete to become fire chief, you have a slight edge. Develop your game plan, make sure your ship is on course, and find your niche to create your own competitive advantage. After all, your success depends upon it!

Attitudes will Dictate Your Altitude

You've heard that "your attitude will dictate your altitude," and I believe that to be true. When I hire new firefighters, the two critical characteristics I look for are not their education and experience, but their attitude and work ethic. Those are the two traits that you cannot teach an individual but that all good employees have in common.

ATTITUDES

The longer I live, the more I realize the impact of attitude on life. Attitude, to me, is more important than facts. It is more important than the past, than education, than money, than circumstances, than failures, than successes, than what other people think or say or do. It is more important than appearance, giftedness or skill. It will make or break a company . . . a church . . . a home. The remarkable thing is we have a choice every day regarding the attitude we will embrace for that day. We cannot change our past . . . we cannot change the fact that other people will act in a certain way. We cannot change the inevitable. The only thing we can do is play on the one string we have, and that is our attitude. . . . I am convinced that life is 10% what happens to me and 90% how I react to it. And so it is with you . . . we are in charge of our Attitudes.

CHARLES SWINDOLL

There exists a synergistic effect that begins with attitude and its relationship to your success.

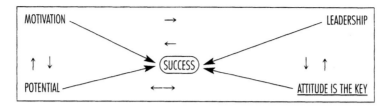

Attitude provides this continual fuel that energizes your motivation, which increases your potential, which promotes your ability to lead, which drives your success. The same holds true as you develop your game plan in the next steps of your career process.

One of the passions that I have to this day is basketball. I still play occasionally and really appreciate watching the game, not only for the athleticism but also for the coaching strategy. When I was growing up, John Wooden was the coach at UCLA. I always admired the way he coached, mentored, and taught his players. He was the consummate professional on the court, and his leadership ability under pressure is unsurpassed.

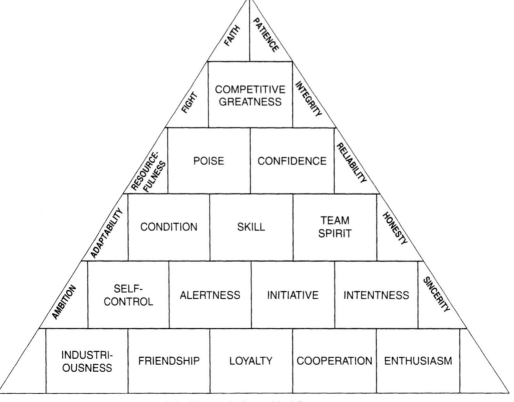

John Wooden's Pyramid of Success

In 1948, while working as an English teacher at Indiana State University, Wooden created his Pyramid of Success. I think it reflects how he coached, how he treated his players, and how he's lived his life. The underlying theory is about trying to be the best that you can be without worrying about others. As Wooden explains his philosophy, it can be transferred to all aspects of life; there's nothing complicated about it. He states "We are not all equal in talent, and all we can do is make the most of what we have and try to improve at all times." I think that is the idea he was able to get across to all his players, and I'm sure that as a teacher he was able to get it across in the classroom.

The Pyramid of Success includes attributes such as friendship and loyalty, self-control, initiative, team spirit, poise, and confidence. At the apex is competitive greatness. That's what we're really speaking about here in regard to developing your game plan for success.

There have been many fine officers who had great potential but did not have the focus, desire, or work ethic and, without a game plan, didn't reach their full potential. I think Wooden's Pyramid of Success captures another, often overlooked, element in why some people are successful and others are not: the environmental impact of having balance. It's a lot like planting a seed in the backyard. If the seed is given plenty of water, good soil, and sun, it doesn't have to try very hard to grow. In fact, there's no way to stop it. If that same seed doesn't get enough water or sun, it will never grow and become a full, healthy-sized plant. It will try because it has the desire to become what it was meant to be. That is an incredibly powerful force, but it will never reach its full potential. In reality, that's the situation for most of us. Whether it's Wooden's Pyramid of Success, your own mission and value statement, or your own code of conduct, you have to develop your game plan, which not only encompasses your professional aspirations but provides a guide to ensure that you are well rounded and have a balanced life. Developing a game plan and making sure that your life is on course is very important. Remember, it's your life; take control of it.

CHAPTER 4

The Blessed Process

You've made the decision to start looking for a fire chief's job. You've talked to your family and you're ready to begin going through the application, pre-screening, and interview processes that ultimately will land you a position. There are certain things that will help you understand the process, your role in it, and the impact on your family.

The Recruiting Process

There are certain issues that you'll find in any recruitment process. One is the difference between a competitive process and a political process. In many situations the search process, even when it is advertised nationally through a recruiting firm, is nothing but window dressing. It is a way to provide the appearance of credibility to the person that the agency has already decided to hire. In many cases, the other candidates brought in for an interview are doing nothing more than providing a façade to validate the process. The appointment was determined before the process ever started.

I've been involved in several recruitment processes for fire chief for different agencies. I have had this experience on two occasions and have found that the knowledge gained does not justify the time spent going through the process. In both cases, the candidate finally selected was in the organization, had minimal education and experience, and was nothing more than a political appointment. While writing this book, I put the theory to the test. I had applied for a job that was rumored to be (and my intelligence confirmed) a "done deal" before it was ever advertised. This job was posted in all the major fire service publications, and I am confident that they received over 100 applications. I had no contact from the agency for over forty-five days until the following letter arrived.

Dear Randy R. Bruegman:

Thank you for applying for the position of Chief, Fire and Rescue Department with _____. The screening committee has completed the interview process and has selected a candidate for the position.

Competition for this position was very intense, with many highly qualified individuals submitting resumes.

Thank you for your interest in this opportunity and for the time and effort you expended in competing for this position.

Sincerely,

Supervisor, Employment
Department of Human Resources

This is a good example of the politics of appointments that occur at times. I think the letter says it all. At least they had the good sense not to bring in several candidates for an assessment interview process. The person who took this position was hand-selected, and the process was nothing more than a façade. That is why, as you go into a job selection process, especially if it is outside of the organization, it is important that you do a good job of gathering intelligence data. One of the best places to start is with the fire chiefs who live in and around the jurisdiction that you'll be testing for. They can provide a great deal of information on the community's politics and its relationship to the organization. If you want a clear perspective on what you are about to undertake, and also gain some insight into the organization, local fire chiefs are an excellent resource. Good intelligence and research will give you a distinct advantage as you go in for your interview. One word of caution—make sure that your interviewees, the local chiefs you have spoken to, are not your competition. As you can imagine, if they are, this may skew the information you get. If the process is truly competitive, then it is up to you to prepare yourself to do your best. It starts with your resume submittal to ensure that you have clearly described your background and experience. It will also help you prepare for the interviews that will occur, an assessment, if utilized, or other testing processes. For example, many assessment processes include an oral interview, a written exercise, a leaderless group discussion, and various presentation formats. In those cases, you need to hone your skills in each of these areas. When making applications, part of your assessment is an attempt to determine what the organization is looking for and match your presentation of experience to its desired qualifications.

What Are They Looking For?

Good questions, and ones that you must ask yourself in every recruitment process, are "Are they looking for someone to come in as a change agent and take the organization in a new direction? Are they looking for someone to maintain the status quo? Are they looking for an outside expert to take the organization to the next level in order to push the envelope to bring about change? Are they trying to fill the position from the inside or the outside? What has been their process in the past in selecting their chief executive and other department heads or appointees? What happened to the last fire chief? Did he or she leave under duress, for a better position, or for retirement? Each scenario offers an insight into where the organization is and also is an indication, if you're selected, of the organizational dynamics that you will face if you take the position. If the former chief left under duress, there may be significant problems that you'll be faced with and will have to address immediately. If you're replacing someone who left for another position but was well respected, you may find that you'll be continually compared to that individual, and you'll have to determine how to make your own mark quickly within the organization to set yourself on a positive track. In many cases, when the fire chief is retiring, you'll find that he or she has actually been retired on the job for several years. In this situation, the firefighters and workforce within the organization may have a great degree of pent-up energy and are waiting for new leadership to move them in a new direction. The reverse can also be found. They may be comfortable with the status quo, because they have been complacent for so long that they have developed inertia, requiring a great deal of energy to get them moving again.

It is also helpful to determine the answers to the following questions. This will help you to assess what your job security will be as you move into this new position.

- What was the job life expectancy of the last fire chief?
- How long has the city or county manager been there?
- Is there a stable political environment or are the politics so extreme that turnover, not only of the fire chief but also of other department heads or the city manager, is extremely high?
- What will your managing authority be?
- Who will you be reporting to?
- Do you report directly to the city or county manager or mayor, or to a board of directors if you're in a district?
- Do you report to an assistant manager or assistant department head? The organizational structure can indicate the importance that the city or county governing body places on your position.
- What will the working relationship be with your new boss? In the interview, ask how your potential new supervisor views the relationship between the fire chief and the city manager or elected official.

When you have the answers to these questions, you can begin to determine your qualifications through your own self-assessment process. There is one final, yet very important, question. Ask yourself "Does my own style of leadership and management make this the job that is right for me?"

Handshakes, Contracts, and Promises

It is interesting that no one contract or working agreement is used for all chief fire officers. In fact, many organizations simply have no contract with their department heads or chief executives. So, it's important to determine early on the common practice of the organization that you're considering joining. Determine your comfort level with that process. In some jurisdictions, the chief executive is appointed with nothing more than a letter of employment outlining the salary and benefits that will be offered. Others have a fully defined working agreement and/or employment contract. If this is the case, your attorney should review it to protect your interests. In either scenario, you need to protect your own interests concerning the little things that often get lost in transferring to a new job and community. This is especially true if you're moving from a different area. Clear articulation of what the organization will offer you to relocate, what that covers, and how it can be used by your family is important. The days of pulling up stakes and moving to a new location on a whim, by and large, are history. There are many reasons. First of all, you and your family may be relocating to a new location. Today there are many two-career families, both spouses with good careers, so career opportunities in the new location may be as important as the job you are evaluating.

There are other factors that must be taken into account today that may not have been important ten or fifteen years ago. The quality of the school system, the safety of the community, and even the overall cost of living can vary dramatically from one community to the next. These differences are much more pronounced today than in the past. The critical factor is, of course, your family. How far will you be moving away from or closer to your extended family? Or is your family already spread across the country, or have you all lived close to each other your entire lives? These are all important factors when you are considering making a move.

To understand your true relocation expense, you will need to compare the cost of living in your present location and the one you're moving to. There are many places where a $20,000 gross pay increase could result in a *loss* of several thousand spendable dollars. Living expenses, taxes, housing, and the overall quality of life must be evaluated in every case. (I know this from personal experience, having moved several times and in each case finding that it took approximately three years to return to our financial position prior to the move.) The main reason for this is the mortgage payment on your house. One of the things that must be taken into account is that when you take a new position and move to a new location, it takes time to sell your existing home. Unless you have preplanned your move very effectively and have already sold your home, you often end up making double payments: on the new home or rental unit and on the home from which you moved. Whether you purchase a home in your new community using a bridge loan or are paying rent, as well as mortgage payments at your original location, it can get expensive. In either case, you have to take this into account when you're thinking about taking a position in another locale. It must be factored in as part of your negotiation process and as part of your preparation to relocate. The more flexible you can be without adding the constraints of a mortgage to deal with, the more options you'll have once you decide to make that transition.

Avoiding the Relocation Blues

There are some steps you can take to help make the transition to a new location much easier. One of the first things to do when considering any position is to call the local chamber of commerce and request information about the area. Today the Internet is an excellent source of information on population, demographics, average income, weather, local hospitals, the health care system, and the cultural and recreational amenities that exist in your new community. Many real estate companies have Web sites on the Internet, useful for comparative analyses prior to your move.

I have noted that one of the most difficult things to assess is what it would really cost to live in the new community. It's not uncommon to find that there are many expenses that you will not have anticipated. Let me give you an example. We moved to California from Colorado. The car we had just purchased did not have a California emissions sticker, so it did not meet California's standard; $900 later, we had license plates. Ouch! The key in preparing to move to a new area, and to make it a positive experience, is to evaluate the expenses adequately and realistically. Do a comparative analysis of such things as housing, a car, insurance, health care, and food costs compared to those at your current location. What are the property and utility taxes, user fees, and special district taxes? All these things can quickly add up and devour even a sizable pay increase. Take time to review your current expenses and investigate what they will be in the new location. Asking your potential employer to provide a sample of your paycheck is a good way to compare the economics picture where you're now living versus that in your new position. Also, call insurance and real estate agents to begin your exploration and collection of comparative data. One of the more frustrating things you can experience is to move into a new area and suddenly find out that your disposable income is less than it was at your old job.

Another critical factor to investigate is the environment in your new community. Population, average income, local race and ethnic relations, community diversity, and crime statistics are all very important. Such information helps to paint a picture of the community. The quality of the school system reflects the community's attitude toward education and their willingness to support it. Again, networking with area fire chiefs and/or the local police chief can be a great help. If you can call a colleague in the area, someone you met at the National Fire Academy or at a seminar, who lives in the immediate area, or a contact through a local professional organization, you can gain insight into what the community is really like. Take the time to drive through the neighborhoods, and look at the quality of housing stock and property maintenance and the quality of life. Consider subscribing to the local daily newspaper, which will tell you a lot about what is occurring in the community and what's important to the people who live there. Remember, many papers today are available on the Internet, even in smaller communities. Don't rely totally on information supplied by any one agency. If you really want to find out about the school system, visit the schools and talk to the staff. Look at the buildings; are they well maintained? Does the school system have adequate capital funding? What's the teacher-to-student ratio? All of these things will reveal the community's commitment to the school system. Here is a tip: The health of the community and the funding levels of local government are indicative of their state of infrastructure. If the roads, parks, sidewalks, and governmental facilities are well maintained, this is a good barometer of their financial stability and of the community's willingness to spend money on basic services.

Preparing Your Move

Many of us have moved across town and found it to be an arduous task. Preparing to move across the state or country is an enormous task, and you must take several things into account to make this move as bearable as possible. Just as preplanning is the key to success in the fire service, it's also the key to a successful move. Not only are you concerned with packing your possessions, loading them on a truck, and relocating to a new community, but you'll also have a host of other duties to handle. Establishing a six- to eight-week calendar before you move and determining what needs to be done each week will make the move much easier. One of the first things you'll have to decide is whether to move yourself or hire a professional mover. In either case, lightening the load by getting rid of a lot of things that you don't use is important. In addition, it is often cheaper to sell the old, save the moving costs, and buy new things when you get to your new location. Begin several weeks in advance by collecting personal records and gathering information on such things as prescriptions and the dates of last examinations. Ask your doctors, dentists, and veterinarians to recommend colleagues in your new community, and obtain the paperwork so that once you find new professionals, they can have your records sent to them. Once you have a new address, you must remember to notify everyone with whom you do business. The list is quite long—utilities, lending agencies, credit card vendors, professional services, post office, motor vehicle division, and others—many of which need up to six weeks' notice to process an address change. Don't forget to prepare a list of your friends and relatives to notify of your new address and the date that you will be moving.

If you decide to move yourself, you need to begin collecting suitable containers and packing materials and setting the deadlines to pack the things in certain areas of your house. It is amazing how long it takes, as the date of your move approaches, you don't want to find yourself panicking and throwing things in boxes at the last minute. Be sure to develop a plan and stick to it. If you're renting a truck from one of the major rental companies, they offer excellent guides to planning your move. These booklets will help you plan the process of packing and loading, as well as select the right type of vehicle you will need to make the move.

In many cases, cities and counties that hire chief executives pay for a professional mover. If they do there are some things to consider. The first is cost. Normally, movers will give you an estimate based upon the total weight that is to be moved. Some, however, provide a binding estimate. A binding estimate, or bidding cost of the service, specifies in advance the mover's precise cost, regardless of the weight you end up moving. Unless you've been given a binding estimate or firm cost, your move's exact cost won't be determined until your property has been loaded onto the truck and weighed.

Here is a piece of advice: Use nationally recognized moving companies. They may be a little more expensive, but their track record on customer service is well proven and documented. A story by Mitch Lipka in the August 2003 issue of *Good Housekeeping* makes this point very well.

Can You Trust Your Mover?

When Janet Smith* and her kids were relocating from Pennsylvania to Texas, the 37-year-old nurse went online and found a "great deal" on a mover. Weeks later, when the truck pulled up to her new home, the movers demanded an extra $4,500 to release her belongings. Smith refused. She called the police, only to be told her problem was a civil (not criminal) matter.

Stacks of complaints like this one led the FBI in March to file fraud and extortion charges against 16 companies, all of which used flashy Web sites to project a professional image. But don't count on much help from the law. (The Interstate Commerce Commission disbanded in 1995, leaving just four federal employees to police the moving industry nationwide.)

Though some of these companies have been shut down, new ones keep cropping up. To avoid their costly scams, consumer advocates advise:

- Forget online quotes—they're virtually worthless.
- Make sure the mover is licensed.
- Have an estimator visit your home; then get several quotes in writing before selecting a mover.
- Ask for references—and make the calls.

Been Scammed? To file a complaint, call 888-368-7238.

* Name changed upon request.

If your employer is paying for the move, it is important to determine the method of payment before your household goods are loaded on the truck. Most companies will gladly take a city or county purchase order. If you're being professionally moved, they can pack the material for an additional price. Whether they do this or you pack yourself, it's very important to work with the mover or local agent to determine how the material must be packed for shipment. If you're doing it yourself, they will supply the boxes and packing material and instructions for you to do it correctly. If the professionals are doing the packing, you'll need to prepare for their arrival one to two days before you actually move, and you'll have to build this time into your schedule. With professional movers, moving day, both loading and unloading, is an experience. The driver of the truck will hire local movers to meet at your house at a designated time. So, just as you do not know them, they often do not know each other. The driver's ability to communicate and direct his or her workforce will dictate how smoothly the move goes. The best advice is to be ready and stay out of their way. Once they arrive, they are on a mission to get the truck loaded and travel to their next stop.

Whether you move yourself or have it done professionally, determine how loss and damage liability will be handled. Insurance can be purchased or, in some cases, your homeowner's insurance will help you cover any loss that occurs while moving your possessions. Become familiar with the limits of the carrier's insurance, be sure you understand the deductible, and note any damage to your belongings before and after the move. One other final tip: allow yourself enough time. If you are relocating out of state, in many cases you will be one of several households whose possessions will be loaded on a semitruck. Subsequently, possessions will be unloaded in accordance with the route to your new home. Therefore, you could be the first off the truck or the last. One of the most difficult tasks is planning when you need to show up to meet the moving truck. Allow yourself enough time to get to your new location.

The Emotional Aspects of Moving

One of the most often overlooked aspects of moving is dealing with the emotions that accompany the move. You will experience a host of emotions during this period, both positive and negative. All must be dealt with if you and your family are to make the transition successful. Your emotional state may be volatile at this point. It's not uncommon to be leaving a tenured position in a community that has become home. Although you are elated and excited to undertake a new position, you may also be grieving for the friends and the community you have to leave. Your family is on that same emotional roller coaster. In today's world with so many two-career families, the career opportunities of your spouse also have to be taken into account. Have you discussed this issue? It is extremely important and should not be overlooked.

In my experience, many emotionally related events do not occur until after the move is made. You, your spouse, and your family are so busy preparing for the move and taking on the new position that you give no thought to your extended family and friends for the sake of your own mental health. When making a transition like this, it is healthier to take the time to feel the loss or regret, and sometimes the fear that goes along with such relocation. I've found that if you don't address it, this emotional event usually occurs about three to five weeks on the new job, when suddenly you realize that your support group of friends and family, in many cases, is no longer there and you're basically starting fresh. That's when the emotional blast hits.

The Children's Side of Things

There is one aspect of moving that I want to stress: the children's perspective. Don't overlook the emotional fallout that your children may experience, which may or may not occur before the move, depending upon their ages. They will experience the same range of emotions that you do—not wanting to leave their friends, being afraid of moving into a new area, being unfamiliar with the school system, and having fears about making new friends. I've found that once you've made the move, it's surprising how adaptable the children are. The younger they are, the more quickly they seem to adjust. That's not to say that they're not going to have emotional ups and downs, but if you're aware of the emotional side of your relocation, you can deal with these issues as they come up. The important thing is to recognize that there is an emotional element to the relocation process. For the children, the sooner you can provide stability and a routine similar to the one they left behind, the better.

The three- to six-month period surrounding the move will be a very interesting and challenging time for the entire family. Understanding the range of emotions that will occur for you and your family will make the relocation easier, and will help you settle more quickly into your new position.

It's true that when you have an unhappy home life, your effectiveness within your organization will also be impacted. So, when moving to a new organization, it's extremely important not only to avoid critical professional mistakes, but also to be aware that the move has significant personal aspects. It's all about balance between your home life and your professional life. If you consider making a move, keep that balance, and the challenges that you face, clearly in mind.

Many chief executives choose to relocate prior to their family to give them time to learn their new position both within the organization and in the community. In many cases, this allows them to focus on helping their family deal with personal issues such as unpacking, living in the new house, and developing a new family routine when they arrive. Others find that having their family with them from the start is more conducive to getting resettled. In either case, the emotional aspects of relocating are as important as the financial aspects and should not be overlooked.

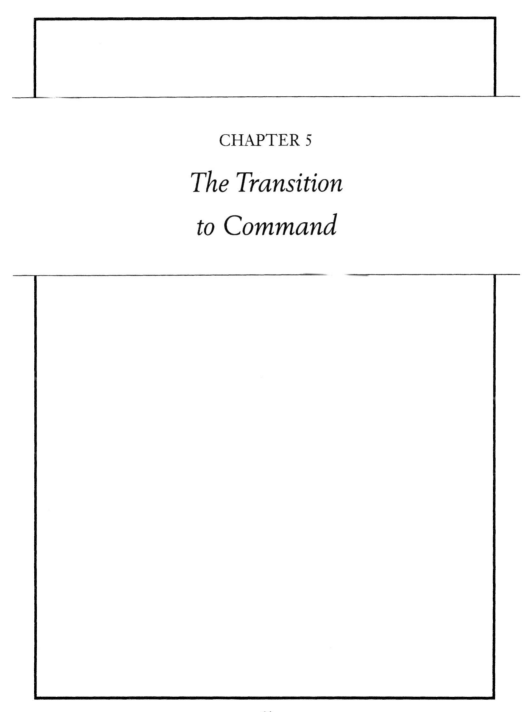

CHAPTER 5

*The Transition
to Command*

The promotion to chief officer is one of the most significant events in your career. Many chiefs have told me that their first gold badge and the job that went with it was the most challenging and rewarding. Through the academic pursuits and experience you have acquired during your career, you now possess the qualifications to take on the role of chief fire officer. However, the transition to chief can be a very difficult adjustment if you have not prepared to take command. Remember, having a degree and many years of experience doesn't always translate into being a successful chief or a successful leader. When I took my first chief's job, I was told on several occasions by my firefighters and company officers that "Now that you're the chief, don't forget where you came from." This is an interesting comment, offering great insight into one of the most significant challenges in the transition to command. That change is reflected in your responsibilities and also in the relationships you'll have with the people you lead. There are many issues in making the transition from battalion chief or deputy chief to fire chief. Let's explore those issues from two vantage points: when you move up within the organization and when you come from outside.

Source: Courtesy of Chief Rob Brown, BCFD.

Change in Relationships

When you are promoted from within, you already have a relationship with the organization because you have come up through the ranks with the individuals you are now leading. They understand that you have some experience and expertise. In many cases you've fought fire and responded on EMS, Hazmat, and rescue calls with many of them. You have lived in the firehouse together, shared the camaraderie, and have been part of the team. However, the relationship changes the day you pin on the badge of chief fire officer. You're no longer one of the group. Probably one of the most overlooked aspects of becoming a chief fire officer is that while *you* haven't changed, your job has; therefore, the way you're perceived by those you lead has changed dramatically. As a chief officer, while you may still be part of some of the social and informal aspects of the organization, in my experience that situation often changes overnight. One of the most difficult things I had to learn when I became fire chief was that my role, and my relationship with my firefighters and in many cases my friends, changed dramatically. This is true whether you rise within the organization or come in from outside. Your role, and the relationship between you and the firefighters, are quite different from those of even a battalion chief or deputy chief. There are several reasons for this:

- You are the chief, so a certain percentage of the workforce will not like you for that reason alone.
- The buck stops with you. As a battalion chief or deputy chief, you can always deflect a small amount of criticism at the fire chief, but when you are the fire chief, you can't.
- Unless you are in a small organization, your interaction with subordinates will be lessened. This tends to create distance, and a lack of trust in some cases. I have found that if you don't see people on a routine basis, it becomes easier for them to think the worst of your actions and to typecast you in a negative way. Remember, it is easier to be negative than positive. This is a prevalent attitude often found in the firehouse, and one that must be understood by any new chief officer.

You have to understand that when you take on the responsibility of fire chief, this problem goes with the job. If you have the illusion that your relationships will not change on the new job, you are in for a rude awakening and probably a short period of tenure on the job.

If you come to the organization from the outside, another interesting situation exists. The people who will be under your command have not worked with you. They don't know your experience and background. They haven't responded on calls with you and often assume that maybe you haven't walked in their shoes. That's one issue you'll have to address early on. Tell them about your background and experience so that they can understand where you have come from and the key experiences and people that have helped to shape your career.

Sharing Your Vision

In either case, whether you have been selected from within or have come from the outside, here are some suggestions as you move into your new role as fire chief. Don't spend too much time remembering where you came from and forget to look where you're going. Becoming accustomed to your new job is like driving a car. If your energy is focused on your rear view mirror (where you've come from), you will undoubtedly run into something because you can't see where you're going. A good chief officer needs to have a clear vision of what the future holds and begin to develop a plan of action on how to take the organization there. Establish your agenda and communicate it. As you are the new fire chief, your personnel are waiting for you to set the tone in the organization. The worst thing you can do is to have a long-term agenda and forget to share it. Hidden agendas tend not to get done, and cause tremendous frustration and stress among your personnel. Communication is the key. Outline your expectations of your personnel and yourself. Whatever they are, writing them down and communicating them will help you to make a smooth transition to your new organization. By outlining your expectations of those you are leading, you will establish a foundation for future success.

Write down the situations, issues, and behaviors that make you react more than others. What bothers one person is often inconsequential to others. As fire chief, you need to let your personnel understand clearly what is acceptable and what is unacceptable in regard to behavior and performance within the organization. This opens the door to frank and honest discussions of areas where personnel and your direct reports may have differing opinions. Determine quickly how you will formalize the process of sharing expectations within your organization. Once you have outlined your agenda and expectations, and have identified for your personnel what your pet peeves are, it is time to communicate your ideas.

Transitioning and Adapting

As you move through this process, take three to six months to evaluate where your organization stands. Determine if your leadership and management style is meshing with those within the organization. Before you took the position, you tried to determine if your style was a match; now that you have done so, this is a good assessment to make. From personal experience, I know that you may have to adapt certain aspects of your leadership style as you move into an organization. I had to do that in two of three moves I made. You can use this as an important growing and learning experience that will continue to stretch your own abilities. Leaders at times need to adapt to the organization as much as the organization needs to adapt to them. Be aware that you may have to do that. Many in leadership roles try to force the organization into their style. (This does not work.) Once you realize that you may have to adapt, it can be an enjoyable and challenging experience personally, as well as professionally rewarding. Remember, this in no way means that you should change your leadership style so much that you become someone you are not. I have witnessed colleagues go through this process, and the end result for both them and the organization was a high level of stress and frustration. Good leaders adapt their style to fit that of the organization, but they should not be chameleons, changing daily.

I speak of this issue as someone coming in from the outside and taking command of four different organizations. You don't want to spend all your energy trying to force your organization to fit your style. When I took my first fire chief's job in Campbell, California, I did exactly this: being inexperienced, I immediately reverted back to my previous style. As soon as I arrived, I eliminated many of the policies and procedures in place and replaced them with those that I had brought with me to make my new department resemble the one I had left in order to make it more comfortable for me. Looking back on that process now, I understand and appreciate the willingness of my staff and the firefighters to indulge me as I learned to become a good fire chief. Subsequently, I took over three other organizations, again coming from the outside. In each case, it became much easier as I became more adept at understanding the organizational dynamics that exist for a chief officer coming from the outside. It's important to understand that in every situation the culture of the firefighters, as well as that of the city and/or county, will dictate how you as a leader adapt in your new role.

The art of leadership involves the ability to blend your strengths with those of the organization. Another element to focus on early in your tenure is to find out where the organizational landmines are. Where do the resentments and hostilities within your organization lie—with the inside person who competed with you for the job and did not get it or with past labor–management issues? Find the landmines and begin to deal with them. If you were chosen from the outside and competed with internal candidates, you need to address their resentment. Clearly outlining your expectations for both your senior staff and yourself can be very helpful in dealing with this resentment and help to alleviate it. Find out what your staff's philosophies are on the critical issues of discipline, personnel management, and the importance they place on the basic tactical objectives of the organization. This will tell you a lot about where the organization has been and where you may need to take it. Determine where the loyalties lie within the organization. There are always hidden relationships that will emerge over the first six months to a year. Loyalties that go back fifteen to twenty years can also create opportunities for you as the chief executive to fail or to experience significant difficulties in your new position. So, it's important to find out where the loyalties are, especially among your senior staff officers. In every department where I have been fire chief, I have found at least one troublemaker in the group. The sad reality is that in every case these people never portrayed themselves that way, and in each case they had to be dealt with. Troublemakers are often found to be in minor conflict with many others in the organization. They pass it off as just part of leadership, as making people accountable. But the truth is that they are using behind-the-scenes, behind-the-back methods of leadership through control and manipulation.

Identify the talents within the organization. Are there resources that are not being utilized, talents that are lying dormant because they haven't had the opportunity to blossom? Look at the level of professionalism throughout the organization. Are many of your firefighters pursuing advanced education or do many of them already have degrees? What is the level of expertise and the educational background of your senior officers? This will tell you a lot about where they have come from, what the mission of the organization has been, and the importance that has been placed on continued professional development. Begin your assessment of the organization with a look at its values and culture. In each of the four organizations I have worked in, the culture and the value system were distinctly different. There are several methods of analysis that you can utilize to gain a perspective on your organization. The one that I have found useful and easy to use, providing a broad view of the organization, is the SWOT assessment. SWOT—Strengths, Weakness, Opportunity, Threats—is a process that you can hand off to each of your division leaders and ask them to assess their own division. This will give you a quick overview of how others view the organization and will allow you to determine where you will need to take the organization in the first twelve months. Each time I have used SWOT, I have found that it also identifies organizational barriers to improvement. For example, if one division leader cites a program as a strength and another leader cites it as a weakness, that signals an opportunity and a barrier—poor communication. The SWOT assessment allows you to open an organizational dialogue on what needs to be focused on and what needs to be built upon.

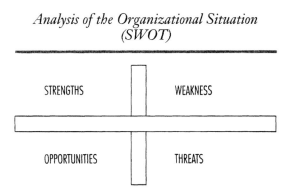

Analysis of the Organizational Situation
(SWOT)

STRENGTHS   WEAKNESS

OPPORTUNITIES   THREATS

Sample SWOT Outline

I. Introduction
Describe the issue being evaluated in detail and explain how it impacts the community.

II. Situational analysis (where are we now?)
SWOT analysis (chart)
Consider the following when preparing SWOT:
A. The Situational Environments
1. Demand trends. (What is the forecast demand for the service? Is it growing or declining? How, when, where, what, and why?)
2. Social and cultural factors.
3. Demographics.
4. Economic and business conditions for services at this time and in the geographical area selected.
5. State of technology for the agency. Is it high-tech state-of-the-art? In short, how is technology affecting service delivery?
6. Politics. Are politics (current or otherwise) in any way affecting the situation for the service?
7. Laws and regulations. (What laws or regulations are having a positive/negative impact?)
B. The Neutral Environments
1. Financial environment. (How does the availability or unavailability of funds affect the situation?)
2. Government environment. (Is current legislative action in the state, federal, or local government likely to affect the agency?)
3. Media environment. (What's happening in the media? Does current publicity favor the agency?)
4. Special interest environment. (Are there direct competitors or any influential groups that are likely to affect the agency?)
C. The Competitive Environments
1. Describe your main competitors, their services, plans, experience, know-how, and financial, human, and capital resources. Do they enjoy the favor of their customers? If so, why? What are their strengths and weaknesses?
D. The Organizational Environments
1. Describe services, experience, know-how, financial, human, and capital resources. Do you enjoy the favor of your customers? If so, why? What are your strengths and weaknesses?

III. The Service Area
Describe your service area in detail by using demographics, psychographics, geographics, lifestyle, or whatever segmentation tool is appropriate.

IV. Problems and Opportunities
State or restate each opportunity and indicate why it is, in fact, an opportunity. State or restate every problem. Indicate what you intend to do about each of them.

V. Objectives and Goals (where do we want to be?)
State precisely the objectives and goals in terms of your strategic plan and the time needed to achieve each of them.

VI. Strategy
Consider alternatives for implementing the overall strategy.

VII. Summary

Take the time to review all standard operating guidelines, personnel policies, and administrative policies of the organizations. You can determine a lot by evaluating how the policies and procedures are written, and the importance placed on each of them is often reflective of the culture and tenor of the organization.

Landmines, Pitfalls, and Red Flags

Your first weeks on the job as the new fire chief are an interesting time, whether you have come from inside or outside the organization. Within the first week, people will be knocking on your door to tell you about all of their good ideas that were squashed in the past. You'll also have a lineup of people telling you the bad things that somebody else has done. While they all may sound valid and plausible, these accounts are a significant landmine for you, because there is usually past history, past baggage, and a hidden motivation attached to each of them. So the rule of thumb here is to listen intently, make no commitments, move slowly, and take the time to make the determination for yourself. That great idea that someone brings forward may have been explored and presented to the city council or the governing body two years ago and been turned down. The worst thing that you can do is to resurrect it and bring it forth as one of your initiatives, only to have it shot down. When this happens, you have sustained an early loss in your tenure. This is often a tactic used by those who want to cause a disruption for the new chief executive. In every department that I've moved to, half a dozen people have told me within the first two weeks all the bad things about someone else within the organization and what I needed to watch out for. In some cases, those insights were valid; in others, they were nothing more than a mechanism of spiteful revenge. As the new chief, you have to be very careful in these situations. As we discussed previously, the relationships within the organization that you're not familiar with can play out in a very negative way if you take the advice of someone you may not be able to trust. So, again, listen intently, take the information, but make the determination for yourself based upon the actions that you witness. I have also found that many "bad apples," as they were labeled, turned out to be some of my best employees.

The Pitfalls of Transition

Don't mistake the six- to twelve-month honeymoon that you will experience as new fire chief for support, because frankly, it is not. In fact, it's a major pitfall for many new chief officers. One of the big mistakes that new chiefs often make is moving too fast. A good chief executive can do a fairly comprehensive analysis in a short period of time and establish a good game plan rather quickly. In every chief's position I have held, I knew within ninety days the major areas that required attention and had a fairly good idea of what steps were needed to fix the problems. I would suggest that you not unveil that strategy for at least four to six months. The organization needs time to adjust to you and you to the organization. Take the time to develop your process of change and decide how you will roll it out within the organization. Begin to develop relationships with your staff and to determine where the landmines are. This can ultimately mean the difference between your success and failure.

From the time we become firefighters/recruits, we are taught to make decisions based upon available information and to do it quickly. This is critical during emergency operations, when it is necessary to evaluate a problem, develop a strategy, and initiate tactical operations—in most cases in less than a couple of minutes. It's natural for us, especially if we have come from an operational position, to carry that decision-making process over to our daily administrative functions. In most cases, however, that's a poor tactic to use when you're fire chief, and especially when you're a new fire chief. One of the attributes that you need to develop in your decision-making process is patience. Patience is often an alien characteristic to many of the action-oriented, judgmental personalities we find in our profession, yet one that is very important to our overall success. The upper level of government, whether local, state, or federal, is as much about process as it is about issues. That is why patience is a virtue. In many cases, your idea, program, or strategy may take several months, or even years, to be fully implemented. Planting the seed of a concept and letting it grow to implementation often requires patience. Timing is everything, and in many cases, knowing when to push and when to hold back is just as important as the issue being discussed.

Another trap for many new chief officers is the failure to foresee the impact of the decisions they make versus those of a lower-ranking officer. As a fire chief, the decisions that you now make affect an entire organization and often the community. Failure to recognize this can cause problems for you, especially when your firefighters, your boss, and elected officials begin to question your decision-making ability. To ensure that you do not fall victim to the "Ready, fire, aim" decision process that many fire chiefs use in their initial transition to command, outline a process and a game plan for yourself on the first six months and stick to them.

The First Ninety Days

Within ninety days of taking command, you probably will have a good idea of the values and culture of the organization. You've identified some areas that need to be addressed and possibly some people who need to be relocated because they're in the wrong positions.

You've done your SWOT analysis, so you know what the strengths of the organization are. You've identified what some of the weaknesses are. You understand the opportunities that exist for the organization and have identified the obvious threats, both internally and externally. Now it's time to begin to lay out your game plan for the organization for the next year. In this situation, I've found that process is everything. Today, with the changing nature of the workforce and the demands placed upon all of our organizations, process can mean as much as the content of the proposal. Therefore, as you plan your rollout strategy, determine what you want to accomplish. Equally important, you will have a better understanding of the organization's culture, which will help your rollout process. The critical factor is how to translate that knowledge into the organizational mission/vision and strategic objectives so that you can begin to move your organization in the direction you want to take it. Start by developing a process by which you can reach out through the organization to include the stakeholders who are critical in making the necessary changes occur. Set an agenda and communicate it. People need to understand what you desire in this process and what the outcomes are for the organization. Trust starts at the top.

It seems like a cliché, but trust is the bond that holds the organization together. It's the emotional response that you and those you work with share. It's a codependency with each other in trying to achieve a common vision, culture, objectives, and all the things that make up the organization. I have attended several management retreats and senior staff workshops held for the purpose of building the team and focusing on common objectives. Some had a long-term, positive impact and others did not. Success depends on a basic element: trust. Get people to agree on a set of common organizational objectives and a joint declaration that they'll all work toward that accomplishment. Bringing a group together, having team-building exercises, and focusing on the common mission and values, is the easy part. The difficulty is in sustaining that effort once you leave the meeting and get back to the day-to-day routines of running the organization. That's where the trust factor is often overlooked. It's when you go back to work and hit that first bump in the road that you begin to measure what the trust level really is. Chief fire officers experience this every day, whether it's with the labor president, volunteer association president, shop steward, mayor, council members, or board chairman. When misunderstandings occur, the level of trust will determine whether you can lead, do collaborative fact finding, explore possible solutions, or move into confrontation. In many cases, due to perceptual biases, the belief is that someone must be doing something wrong, hiding something, lying, or just plain ignorant. Every one of us has received a phone call in which the person at the other end is presuming guilt. "What are you guys doing over there?" "When are you going to fix the problem?" "Hey, it wasn't my idea that those guys were going to do that." We've all gotten those calls, and when we do, most of us tend to become defensive. When you do that you often go offensive, as a means of protection to level the playing field. With a higher degree of trust, that interaction is different. So is the outcome. The pattern seems to be: low trust, high negativity; high trust, high positive outcome. This correlation is not a fluke.

As leaders, we need to have a high level of trust in our relationships with key individuals; otherwise, we reduce our ability to focus our energy and attention on the right things. Consider the following examples showing the importance of trust. You and your labor leader are dealing with a delicate and serious personnel issue. Concurrence on process and the level of discipline has been agreed on, when one of the parties at the last minute changes direction. When this happens, the trust between the individuals involved, the employees, and the overall labor–management relationship is at risk. The damage is done, and no matter how many positive experiences you may have had, it only takes a few negative ones to do permanent damage. In another example, your boss directs you to implement a contract. When it comes to light that your boss did not follow the contracting rules, he or she denies having told you to move ahead. What is the level of trust now? As a final example, you find that one of your senior chief officers is not performing his work. His time is suspect and he takes credit for the productivity of the division while not contributing to it, yet he portrays himself as the motivating force. Is your trust in that individual damaged?

What we're speaking about here is how conflict is often created. I'm sure you can identify people to whom you may react differently because of past events and the low level of trust that exists between you. Your barrier is higher, and you are less willing to listen to those persons than to someone else who comes to you with a problem or an issue. I have had a tendency to do this, and it has forced me to step back and listen before reacting and evaluating before making a judgment. Remember those decisive, critical, judgmental people I spoke of previously? They're back! And they (you) can have a tremendous impact on these interactions and the personality they bring to the table, so, let me offer some tips for overcoming conflict:

- Listen so that people will talk and talk so that people will listen.
- Focus on the problem, not the person.
- Build "power with others" not by "powering over others."
- Express feeling without blaming others.
- Own your part of the conflict.
- Strategize to reach mutually agreeable solutions.
- Create options: one-way solutions always create losers.
- Solve the problem and build the relationship.

These tips have proven to be very useful for me and I hope that they will be for you as well.

I also find it interesting that in participating in some of these discussions and watching others, we often become polarized. This is not because of the issues at hand but because of past issues in which our interactions with particular individuals have polarized us in regard to the issue of the day. If I find myself having a difficult interaction repeatedly with a specific individual, I often go back and review these tips on conflict, jog my memory, and, in some cases, remind myself to play well with others.

Too often we let our thinking and our beliefs about what we "know" prevent us from seeing things as they really are.

. . . [A]n open "beginner's" mind allows us to be receptive to new possibilities and prevents us from getting stuck in the rut of own expertise which often thinks it knows more than it does.

JON KABAT-ZINN
UNIVERSITY OF MASSACHUSETTS
FACULTY C. 1946

The relationship between the level of trust and conflict has to do with several factors outlined in the following table. The resulting interaction can range from war to problem solving.

THE LEVEL OF CONFLICT IS OFTEN DICTATED BY THE LEVEL OF TRUST

The Motivation	The Objective	The Assumption	The Emotional Climate	The Communication Style
Problem solving	Solve the problem	We can work it out	Hope	Open, direct, clear and nondistorted; common interests recognized
Disagreement	Self-protection	Compromise is necessary	Uncertainty	Cautious sharing; vague and general language; calculated "thinking"
Contest	Winning	Not enough resources to go around	Frustration and resentment	Strategic manipulation; distorted communication; personal attacks begin; no one wants to be first to change
Fight	Hurting the other	The other side cannot or will not change; no change necessary on our side	Antagonism and alienation	Verbal/nonverbal incongruity; blame; perceptual distortions; refusal to take responsibility
War	Eliminating the other	The costs of withdrawal are greater than the costs of staying	Hopelessness and revenge	Emotional volatility; no clear understanding of issues; self-righteousness; compulsiveness; inability to disengage

The level of trust often dictates where you start in dealing with the problem. If you have a high level of trust, you are probably much more attuned to solving the problem versus going to war. In our leadership roles, it is important to make sure that our people are focused on the right things and not on the people involved.

Leaders have a harder job to do than to just choose sides; they must bring the sides together. Choosing sides may not always be issue-related; it may be about how leaders perceive themselves. When we move up through the ranks, we tend to become less relationship-oriented and more outcome-oriented. It becomes easy to focus on evaluating organizational performance. But if we truly wish to move the organization forward and increase its performance, we can't forget about the emotional glue that holds it together and the processes that we use to move ahead. Trust doesn't always guarantee that the road we have to travel will be easy, but I can assure you that it will provide a smoother ride when conflicts erupt and barriers emerge. Trust in organizational relationships often leads to greater understanding and a willingness to seek constructive solutions instead of becoming defensive and overprotective. As a chief fire officer, you have to earn this trust. It can't be negotiated, hired, or tacked on to a paycheck. Your actions will often determine the road that your organization will travel. There are actions and behaviors that reduce and build trust.

Actions and Behaviors That Reduce Trust

- You act more concerned about your own welfare than about anything else.
- You send mixed messages, so people don't know where you stand.
- You do not take responsibility for your own actions. You pass the buck or drop the ball on projects.
- You jump to conclusions and react without first checking the facts.
- You tend to make excuses or blame others when things do not work out.

Actions and Behaviors That Build Trust

- You communicate openly and honestly without distorting the information.
- You show confidence in the ability of your personnel and treat them as skilled, competent associates, not subordinates.
- You listen to and value what others have to say, even when you don't agree. You're inclusive.
- You keep your promises and commitments.
- You look for ways to cooperate, collaborate, and provide opportunities to help others.

In his book *Have Trust in Partners*, Jordan Merris provides insight into six conditions for trust in most organizational relationships.

- Each mutual need creates opportunity. When a mutual need exists, it motivates people to be more open and commit the necessary energy to be successful.
- Interpersonal relationships make the connection.
- The encouragement of successful alliances begins with people. When you can bring the project or objectives alive through the people who are involved, trust is built. This creates value and allows those involved to tackle even the most difficult issues.
- Empowered leadership. This occurs when your management staff is empowered, free to roam, and running on the same track, which sends a clear and decisive message about the direction of your organization and the culture that you are trying to develop. By contrast, polarization at the top virtually ensures that you'll experience conflict and will not solve your organizational problems.
- Shared objectives guide performance. Are organizational objectives mutually agreed on? For example, are your strategic plans or documents created by those who share the objectives and actually have to carry out the plan, or do senior leaders produce them and send them out, thinking that others will follow? A sense of ownership in the plan is crucial to creating trust in the organization.
- Safeguards encourage sharing. Are your people allowed to fail, and if so, how does the organization react? Gaining trust is also about creating a safety net so that people are willing to engage and take risks. I'm talking not about those that will break the organization's rules but rather about encouraging people to propose new ideas, concepts, and procedures that will help carry the organization into the future.

Leadership/Communication

To build trust in any organization, the key is communication. In today's ever-changing environment, trust is even more important. Marilyn Moats Kennedy, founder and managing partner of Career Strategies, based in Wilmett, Illinois, offers the following advice for uncertain times:

- Let employees know that they matter.
- Minimize employee disappointment. Be honest; don't pretend that times are not tough when they are.
- Overcommunicate—have frequent, direct, and honest communication with your employees about what is going on in your organization.

- Cross-train and work in teams. Then, when someone leaves, those who remain can cope.
- Don't punish cynics. Cynicism and negativity have no place in public, but recognize that cynics often have more realistic expectations of the organization.

The bottom line is that people respond best to the personal touch, especially when conditions are unstable. The above is good advice if we are to build trusting organizations through open communication, cooperation, and inclusiveness.

Key questions you should ask when moving up or into a new organization

1. What is the level of trust in the organization?
2. How willing are personnel to be honest with each other?
3. Does the organization have the ability to communicate up, down, and sideways?
4. Does the organization kill the messenger?
5. Is the focus on problem solving or decision making?
6. To what degree do personnel share values? Do they all have the same vision, focus, and goals?

These are important questions as you try to determine the problems that exist, the barriers, and what approach(es) or action(s) will be needed to establish effective communication.

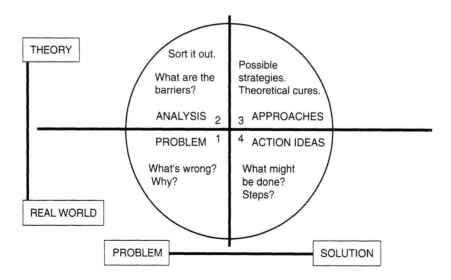

A critical element in developing a high level of trust and getting people engaged is motivation.

How can we "fire people up"?

1. Do you intend to be creative?
2. Do you aim to use creativity to solve problems?
3. What, if any, demands does the organization make on its personnel?
4. Are small problems more important than big ones?
5. Do you fully understand a problem first and then solve it?
6. Is value placed on options or just solutions?
7. What is considered acceptable failure?

From a leadership perspective, as we institute a higher level of trust within the organization, we can focus more of our energy on the opportunities for change instead of overcoming organizational constraints and barriers.

A higher level of trust will help drive needed changes.

As people begin to trust one another more, their commitment will increase and their best efforts will be given to the work. Interestingly, even if the project or task is a failure, the commitment and the shared experience will carry over to the next project and the next issue and will ultimately have a positive impact on the organization.

Organizations with good internal collaboration often have better external relationships as well—with the public, vendors, and other governmental agencies. Part of your leadership role will be to determine where your organization stands. If divisions and departments are busy bickering among themselves, they won't have time to build good external relationships on behalf of the organization. In addition, they won't have time to focus on the issues within their own divisions and departments, and an inordinate amount of time will be spent worrying about what everyone else is doing. Organizational trust is all about a shared alliance among people. As people move on, take different jobs, and retire, maintaining the attitudes and understandings that have been developed over time is critical for an organization. This intangible factor cannot be measured, but it is a critical component of successful organizations.

I've found an interesting dichotomy in the fire service. In our job, on a day-to-day basis, our trust and our very well-being are in the hands of others as we deal with crisis situations. Yet when we return to the firehouse, we become much less trusting toward one another and toward others within the organization. That's a major challenge for leaders in the fire service today. We must take advantage of the opportunities to build trusting relationships. We must also understand our own weaknesses and shortcomings, and the processes of dealing with different personalities daily. We must understand that we can't get bogged down in the minutiae of rhetoric and politics, but instead must stay focused on the mission of our organization and our service to our constituents. We must also make a commitment to work on those relationships within our organization that are often negative and nonproductive. True leadership is about the ability to recognize weakness, whether it resides in us or in our organization, and to tackle it head on. It's about tapping into the leadership potential throughout the organization and developing it. Trust in your organizational relationships and in the individuals you lead will dictate how successful you ultimately will be. Here is another perspective based on my own experience. As a new chief fire officer, you probably believe that leadership is about the ability to control things. Therefore, you may want to be involved in all of the significant projects, go to all of the meetings, and read all of the documents. As you gain more experience and understanding, you realize that leadership and managerial control are not about those things at all. They're about being able to connect with your people and move them toward a common vision and a common mission. I believe that the sooner you can understand and become comfortable with that concept, the more successful you will be as chief fire officer.

Unfortunately, some chiefs never learn this and continue to struggle throughout their professional lives. In many cases, they just can't let go. They read all the documents and sign off on everything that happens in the organization. They are consumed by "being in charge and being kept in the loop." These are also the people who are in a constant state of frustration and dissatisfaction.

Other chiefs grasp the concept that working to achieve a common vision is what will make them very successful at what they do. Leadership and managerial control concern the processes you use to achieve that vision—processes that are good for your people and for you as the leader!

You have to create a process by which the people you command have the opportunity to become stakeholders in the issues that impact them and the organization. By blending their ideas with your vision of the organization and incorporating them into your long-term plan, you create ownership. Once you have developed that process, you must communicate it effectively to all of your personnel. There are several methods you can use to do this. Updating your organization's strategic plan is a means to incorporate all employees' ideas and vision into the organization. Remember, it will be the stakeholders, both internal and external to the organization, who will help drive your proposed changes, not only within the organization but within the community as well.

You can use the quality processes that are available today to set up teams to address specific issues within the organization. During the past two decades, many communities have experienced significant growth and an increased demand for fire service. The fire service has begun to develop a corporate culture that maximizes the participation of employees through the use of committees and/or quality process action teams. The need to empower personnel at all levels within an organization is no longer the latest management fad but is evidence of the cultural shift within the fire service. Quality improvement processes have proven to be effective. The critical point is to understand that whatever device you choose to use and are comfortable with, the process you establish is often as important as the agenda you set. Without a good process, you may not get the buy-in that you need to make the necessary organizational changes.

Don't forget to get out into the community once you have established this process. One of the things we often forget about in the fire service is the importance of community relations. I'm not talking about public education and information. I mean telling the community what your organization is about and what your vision of the organization is for the future. It is amazing what happens when people understand where your organization now stands and where you wish it to go. Creating stakeholder relationships throughout the community is critical during your first year if you're to be successful over the long term.

Last, but not least, don't overlook your leadership team. Depending upon the size and makeup of your organization, your leadership team can be very small or very large. These are the senior chief officers whom you will rely on heavily in your tenure as fire chief. You need to look at the atmosphere in which they work and determine what they expect of you. This will tell you a lot about what they deem important, and it will help you understand how to develop a positive atmosphere. At the senior officer level, it will take a conserted effort to develop a team and foster a "win-win" attitude within the group. It's extremely important to spend time and energy developing a positive relationship with your senior staff executives. Depending upon the situation within the organization, this may take very little energy or it may require most of your time the first year on the job. Don't overlook its importance. Remember, these are the people who will make or break the policies, procedures, and strategic direction you hope to forge for the organization. Your team is a critical component of your success; they will need your attention if you expect to build a quality relationship with them. In some cases, this means that people will have to be moved between positions and off the senior leadership team. Remember, you can't win the Super Bowl with all eighth-round draft choices and players with bad attitudes. If you are to develop a winning team, you need a balance of talent and heart!

We often overlook one of the most important relationships that we can have as chief fire officer, and that is with our counterparts in public safety—police and EMS services. Take the time to develop a good working relationship with your local police chief and/or sheriff. That relationship will pay many dividends in the future. Also, begin to establish a good relationship with your local media. Often in my experience, a potentially negative story played out well in public because of good media relations. The media can also provide important insights into the community and be good sources of information on how your organization is perceived by the community. But you will have the opportunity to gain this insight only if your relationship with the media is good.

In the fire service, our perception of public information is often limited. We wait for the media to call us after an incident occurs. While this is an effective way to provide exposure for the organization, an update to the public on the circumstances of the day, the exposure is limited. Media relations must go far beyond that. We must become a quality resource for them, and the best way to do that is to provide them with information. The media are driven by breaking news, but they also have slow times and are continually looking for educational pieces and "feel good" stories to provide a balance to the day-to-day reporting. The fire service is loaded with this type of information. If you want to create a good relationship with the media, keep pitching stories to them, be available when they call, and establish a relationship with the reporters and producers. You will be amazed at what can be accomplished.

Last, but certainly not least, look at the position of the organization within the community. How does the general public perceive your organization? Almost all fire departments in this country are viewed by the public as expert professional organizations. Fire services start with a 90 percent approval rating. We are well respected and trusted. However, this does not automatically mean that we are doing a good job. As chief fire officers, we have to be willing to look past the high scores on public surveys as the sole measure of our performance. The true test of how the community feels about the fire department may not lie in what you do from an emergency standpoint (after all, the expectation is that when they dial 911, you will respond and do a good job). In most cases you do. Let's be honest. How would they evaluate your performance anyway? Many of us had the experience of a firefighting event that did not go perfectly, yet the next day the homeowner showed up at the firehouse with a plate of cookies and thanks for your help. Don't interpret their view of our profession as a measure of positive organizational performance. I have observed many fire departments that were less than stellar performers but were loved by their community. Just as important as the community's perception of your department's performance is another factor that must be evaluated and nurtured: your involvement in other aspects of community life.

Operation Santa Claus

A wonderful example of community outreach and involvement involves a program that I inherited as fire chief at Clackamas County Fire District #1 in Oregon. It's a program called Operation Santa Claus. This program began in

Source: Courtesy of Clackamas County Fire District #1.

Source: Courtesy of Clackamas County Fire District #1.

the 1980s when a group of volunteer firefighters were contacted by a family in need that lived near one of the firehouses. They took up a collection of food, clothing, and money that ensured that the family had something for Christmas. What started with outreach to one family evolved into a program of seventeen nights when fire vehicles, with lights flashing and sirens sounding, go through different areas of the community with Santa in tow. An old fire truck is decorated with Christmas lights, a sound system, and a sleigh in the hose bed for Santa and Mrs. Claus, and the magic begins. Other fire equipment follows to collect food, money, and toys for those who are less fortunate. The fire

Source: Courtesy of Clackamas County Fire District #1.

department's organizational commitment provides food and toys for Christmas for over 400 needy families within its jurisdiction. The other aspect of Operation Santa Claus is the involvement of the community. The local chamber of commerce holds a Christmas breakfast and is involved throughout the collection and distribution process, including the local regional shopping mall that puts up a Giving Tree where all the toys for Operation Santa Claus go. The local Rotary Club annually holds a special night for underprivileged children in the local school district, serving almost 300 children who otherwise would not receive any toys for Christmas.

Operation Santa Claus is a fine example of how a program, as it grows and is nurtured, can create a mechanism for extensive community involvement. I know from personal experience that it has changed the way the community looks at and reacts to this particular fire department.

The degree of involvement, or lack of involvement, will tell you a great deal about the position of the organization within the community. It should be a factor as you develop your game plan on how to foster a better relationship and stakeholder involvement within your community. It is often overlooked, but within the first few months of taking on your new position, you will learn how important it can be.

CHAPTER 6

Ethics, Politics,
and Leadership

The issue of politics and ethics is one of the most difficult leadership challenges you will face as fire chief. Joann B. Seghini, Ph.D., of Midvale City, Utah, describes what she, as mayor, sees as the fire chief's role and responsibility. It is a fine overview, but it does not address an important issue that fire chiefs have to deal with daily: politics and ethics. The political process is an aspect of our job that is vital to the success of our organization, yet it is one that most of us learn only by on-the-job training. This process, if not understood, can create enormous frustration and can be one of our most difficult challenges. It's one aspect of the job on which we don't receive a great deal of academic background, experience, or training prior to accepting the job as fire chief. In addition, politics has a negative connotation for most people. Nevertheless, it is a real and an essential fact of life, especially for those of us who are fire chiefs. As a new chief, understanding the political landscape may determine the posture and position you take throughout your period of tenure.

The Fire Chief: a Mayor's View

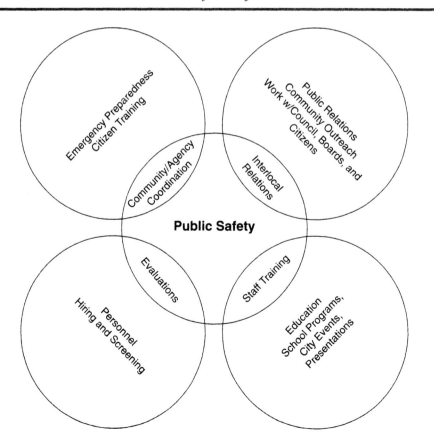

Many don't understand that and feel that they're in a continual state of chaos. The political landscape that we operate in is often very rough and can change very quickly. Sometimes we have fire personnel who think of themselves as politicians, to the point where it is sometimes unclear whether their role is to put out fires or to be on the phone with elected officials. On the other side, we have elected officials who fancy themselves as would-be firefighters. They long to take a more hands-on approach to the operation. The fire chief often feels caught in the middle, and this creates fear. This is coupled with the fact that many people who become fire chiefs have had very little experience with politics and do not understand how the political process works. This can create a very frustrating situation for many new chief officers. We must learn that politics and political gamesmanship is a way of life for most fire chiefs today.

It is important to consider how we integrate politics into our organization and how the political structure integrates fire chiefs into the dealings of the government that they work for. There is something that, as a new fire chief, you have to come to grips with early in your career. I call it "political ethics," which means being clear about what you will and won't do in terms of organizational politics. What is acceptable to you as a fire chief? Is it acceptable for your subordinates to deal with elected officials behind your back? Are officials going directly to your subordinates without your knowledge or is there a strict chain of command that is followed by both? Whatever the case, you as fire chief need to understand what the ground rules are. Having worked in a number of organizations, I know that organizational politics are often as different as people. What is acceptable in one organization is not tolerated in another. In some organizations, department heads and staff routinely talk to elected officials and vice versa. In others, everything goes through the manager or mayor. Understanding the political dynamics within an organization is critical, especially if you're the new fire chief. Determining what your comfort level is and what you are willing to accept, understanding the rules of the game, and identifying the key players in terms of roles, relationships, and responsibility is vital.

The Past Is a Window on the Future

One of the best ways to understand the political dynamics of any organization is to understand what occurred in the past. It's important to acknowledge that there were people in your position long before you arrived. The reality is that much of what occurs today is built upon events or relationships that occurred in the past—some good, some bad. Past organizational events or relationships often dictate the present political structure and often impact the current situation. It's important to understand that structure and those relationships as quickly as possible. Organizational history is a window on the future when you face issues similar to those of your predecessors. The reactions and perceptions will often occur the same way, and it can be extremely helpful to understand and determine how to approach and present an issue or, in some cases, avoid it. Know what your principles are and live by them. This issue concerns the personal values that you bring to the job, and it is very important to deal with it.

This chapter is titled "Ethics, Politics, and Leadership" for a reason. In these three areas that continually cross in the role of fire chief, the chief requires a framework of principles within which to operate. When we speak of ethics, we speak of the principles that help to define proper behavior. Such principles don't always dictate a single course of action but instead provide a "measuring stick." In choosing between different options, it is an important element in being a good public servant and an effective fire chief. Making good ethical decisions often requires the ability to make distinctions between competing choices. Many of those choices involve those three areas for fire chiefs that frequently overlap and require that our integrity be our best asset.

I've witnessed many fire chiefs cross the ethical line to maintain their position, enhance their political stock, or overlook questionable processes to maintain political harmony. It's important for you, as a chief fire officer, to have an ethical guidepost. The IAFC has adopted a Code of Ethics that I have used on many occasions when faced with competing choices. It has provided a touchstone for issues that have forced me to challenge my ethical boundaries. These ethical principles are consistent with professional behavior of chief fire officers, and they include the following:

- *Recognize that we serve in a position of public trust that imposes responsibility to use public[ly] owned resources effectively and judiciously.*
- *Do not use a public position to obtain advantages or favors for friends, family, personal business ventures or for ourselves.*
- *Use information gained from our positions only for the benefit of those we are entrusted to serve.*
- *Conduct our personal affairs in such a manner that we cannot be improperly influenced in the performance of our duty.*
- *Avoid situations whereby our decisions or influence may have an impact on personal financial interest.*
- *Seek no favor and accept no form of personal reward for influence or for official action.*
- *Engage in no outside employment or professional activities that may impair, or appear to impair, our primary responsibility as chief fire officials.*
- *Comply with all the laws and campaign rules in supporting political candidates and engaging in political activities that may impair professional performance.*
- *Handle all personnel manners on the basis of merit.*
- *Carry out our policies that are established by our elected officials and policy makers to the best of our abilities.*
- *Refrain from financial investments or business that conflicts with or is enhanced by our official position.*

- *Refrain from endorsing commercial products through quotations and use of photographs and testimonials for personal gain.*
- *Develop job descriptions and guidelines at the local level to produce behaviors [in keeping] with the Code of Ethics.*
- *Conduct training at the local level to inform and educate personnel about ethical conduct policies and procedures.*
- *Have systems in place to resolve ethical issues.*
- *Orient new employees to the organization's ethics program during new employee orientation.*
- *Review the ethics management program in management training experiences.*
- *Deliver accurate and timely information to the public and to elected policy makers to use when deciding critical issues.*

The ethics of leadership that you use will define you as a person. In the past several years, we have witnessed so many ethical misadventures of those in leadership roles that they have become commonplace, from fabrication of resumes to altering response data to meet the adopted local standard. What always amazes me in these situations is that when these people are exposed, they deny their actions or, in many cases, attempt to blame them on someone else. This is a comment on the integrity of the individual but also on the integrity of the system he or she works in. The individual, the system, and the industry as a whole are tainted by the experience.

The Political Dynamic

The credibility you establish at the local level and the integrity you display will determine your ability to have an effective relationship with your elected officials. That alone will not be enough, however. When you present an item for consideration to an elected body, not only will it be judged, hopefully on its merit, but in many cases it will be judged politically as well. Understanding the political nature of the fire chief's position and how we interact within it is very important. Many of my proposals were the right thing to do based upon good science and yet they were not acted upon, not because of the objective evaluation but because of the politics involved. As fire administrators, most of us have come through a typical career model of being evaluated and promoted based largely upon our training in emergencies. This model is fairly linear, fairly black and white; something is either right or wrong. The model is fairly objective, and we tend to like it. It's easy for us to make a plan, set an objective, develop a course and take action, reevaluate, and move on. The political arena, however, is very different. Elected officials often lean more toward the subjective or softer side of the issues. This is often a key element in their process of decision making, and fire chiefs must be able to understand it and learn to accept it. Understanding the reality of politics and the impact on you as chief is extremely important. When you lose a battle because of a political versus an issue-based decision, you cannot take it personally; you have to be able to let it go and move on to the next issue. Understanding that this is part of the process, and not a personal attack on you, will let you live to fight another battle another day.

Survival, though, depends upon how we get through those loaded political encounters. It's very important to continue to remind yourself of what the outcomes and the objectives are, and what you're trying to achieve in the political process. Remain focused on the issue and not on the people involved in the debate. If you do not, you will begin to focus on the personalities and the political discussions taking place and lose perspective as fire chief. I've seen many proposals voted down not because there was a lack of political support, but because the person presenting the issues mistook the tough questioning as a personal attack. In one case, instead of listening, the fire chief mounted a full assault on the council member, only to remember too late that fire chiefs and senior staff members don't vote. In this scenario you can become very vulnerable and lose your own credibility—and, if it is severe enough, your job. You must always remain focused on the issue at hand and remember why you presented it in the first place. Remind yourself to listen to all the options that are stated. Though you are the fire chief, there are times when politically motivated suggestions are thrown out. Some of them might be good ideas, and you have to be willing to consider them. Getting through politically loaded encounters requires a fire chief to be focused yet at the same time have some flexibility. Be politically sensitive and open to alternatives.

How can you help to create a positive political climate for yourself? The more you can deal with people face to face, the better off you'll be. Relationships are established best with one-on-one encounters. Increase your own tolerance for ambiguity and, when dealing with elected officials or a political body, remember that they may not be used to handling the black and white issues of the fire service. Ambiguity is sometimes the saving grace for politicians, and they often choose it over the more definitive proposal.

Become more receptive to politicians' ideas as they relate to your proposals. Take an open, receptive stance when you bring issues forward. Take a broad view, one that not only incorporates the needs of your organization but also takes into account the city's or county's goals and objectives. Understand that buying a new ladder truck might not be the most important thing on the agenda of the city or county at a particular point in time. These officials may have more pressing needs or issues to address. Share your own goals with your elected body; hopefully, with understanding, they'll begin to support what you're trying to do organizationally, both on a short- and a long-term basis. If they don't, it's a good indication that either they are out of step with the goals and objectives of the organization, or that they don't understand, or in some cases that your credibility is so low that you need to look for a new job.

Remember that understanding does not always mean agreement. Understanding the political nature of the community and how it operates is very important to most fire chiefs' survival. That doesn't mean you have to agree to the final decision on an issue. What it does mean is that you need to understand why the decision was made and see that it is carried out. Sometimes a decision will be reached for totally political reasons that probably go against the grain of most fire chiefs, but this is the reality that most of us are faced with. In either case, whether you agree with the decision or not, it is your job to support it.

The late president of France, Charles DeGaulle, once said that politics is too serious to be left to the politicians. In local government, that is certainly the case. Politics in itself is neither good nor bad. It largely depends on how it's used, the people involved, and their own motivations. The political climate that exists in a given community will create an environment of either trust, distrust, or skepticism. Strong loyalties can form and strange bedfellows made as a result of the political dynamic. All, to some degree, will influence how the political process will work within and/or upon the organization. That's a reality you should not ignore if you wish to be effective.

The Ten Commandments of Political Survival

Several years ago, George Protopapas published his version of the Ten Commandments of political engineering. I'd like to paraphrase them in regard to the role of fire chief. The Ten Commandments of Fire Chief Survival, some of which we have touched on and others not, provide a good road map that allows us to deal with the politics of being fire chief today.

1. Never show animosity toward any elected official. Sometimes elected officials will do or say something to make us look bad in public, either at a meeting with elected officials or in front of a constituent. Always be professional. Coolness under fire can go a long way toward creating a good relationship with the entire elected body.

2. Know your budget thoroughly. Elected officials often refer back to specific areas of the budget that you may not have looked at for the past several months. Questions on your budget often come when you least expect them. When you're presenting an item that has a big budget, has no cost ramifications, and has already been approved, you may be asked questions about something totally unrelated and you are expected to have an answer. So, it is extremely important that you and your staff have a good understanding of everything in your budget and where it can be found at a moment's notice.

3. If you have a proposal but know you don't have the votes to win, don't bring it to the governing body. Give yourself time to lay the groundwork before you place it on the agenda.

4. Stay out of political campaigns, especially in elections that will result in replacing people on your elected body. It's a lose-lose situation. If the incumbent you choose to support loses the election, you will have an opposing vote on many of your proposals from the new elected official. Conversely, if you support local candidates, they often think that you owe them a special favor, which can present ethical dilemmas later.

5. Make a point of getting to know your elected officials better. When you're attending conferences or other events together, get to know them personally. This can tell you a lot about what motivates them, what's important to them as individuals and not as council or board members.

6. Encourage your elected officials, city and county manager, to participate with you and your personnel in events if that is allowed by the political structure of your governing body. It's a great opportunity for them to learn more about the organization and its people and provides a way to share information.

7. Conduct field trips for your elected officials. In the fire service, we have a lot of impressive equipment and great people. A good way for us to create a positive relationship with our elected officials is to encourage them to spend some time with us doing what we do. Allowing them to ride along and participate with the organization in prearranged events will help these officials see what the fire service does on a day-to-day basis. This can offer a great opportunity for discussing with them issues of importance to the organization. Remember, once is not enough.

8. Follow up every problem referred to you by an elected official, even if it is a minor one. Do this immediately, return phone calls, and write to the official to confirm that the matter has been handled. Deal the issue in concert with your city and county manager, if appropriate.

9. Work with your elected officials individually before the open business session. Often, being able to field questions from individual elected officials helps them to understand more fully the issue that you are bringing forward. It can also be a great opportunity for you to learn what they will be asking when the regular open business meeting begins. If you do this, you can't do it for just one official; you must do it for all. Make sure that you don't offend any board members by leaving them out. Again, this has to be done in concert with, and with the approval of, your city and county managers if appropriate.

10. Sometimes your recommendations are not going to pass; you must recognize when that occurs. When it does, don't continue to push the agenda item and end up antagonizing your elected officials. Let it go and move on.

Understanding the methods of political survival for fire chiefs and chief fire officers today is essential. We've talked a great deal about ethics and understanding the political dynamic of our organizations today. Leadership, ethics, and politics are interwoven into the fabric we call government. As chief fire officers, we have to be dedicated to the concepts of effective and democratic forms of government in our response to elected officials and professional management. This is essential to achieving the objective of providing quality service. To do so, we need to maintain a constructive, creative, and practical attitude toward the way local government works. We also must understand and believe that each of us in this role has a responsibility as a public servant. We can fulfill that responsibility only by being dedicated to the highest ideals of honor and integrity in our public, personal, and professional relationships. Our measure of success is the respect and confidence given to us by our elected official, our peers, our employees, and the public.

While chief at Clackamas Fire District #1, I developed a code of conduct. It's something I have tried to live by, and I offer it to you as a starting point as you deal with the political and ethical issues as a chief fire officer:

* Do no harm.
* Place the safety and welfare of others above all concerns.
* Seek to help those who are in difficult circumstances.
* Provide service fairly and equitably to everyone.
* Assist and encourage those who are trying to better themselves in the fire service.
* Foster creativity and be open to the innovations that may improve the performance of our duties.
* Comply with the laws of local, state, and federal governments.
* Defend the Constitution.

Probably no single action that we as leaders can take will affect our credibility more than ensuring that our behavior is ethical and professional.

Scoring Political Points

The fire chief's role has changed dramatically over the past twenty-five years. The biggest shift may have been in the political arena. The fire chief was once the top firefighter; today the chief is a CEO, content expert, facilitator, and politician. While not elected, the position of fire chief stands out and can become highly politicized very quickly.

The fire chief today must understand as much about the political process as he or she does about incident command. Whether you run a small all-volunteer organization or a large career metropolitan one, you will work in a political environment. It may involve the volunteer association, the local union, citizen groups, and/or elected officials. You must have the expertise to blend your skills, ethics, values, and organizational needs into the political process. Equally important, you must blend these in such a manner that you are professionally successful, create organizational improvements, and have positive relationships throughout the political arena.

There are some guidelines for political survival for the fire chief today that are essential to our long-term job success and, in some cases, our survival. These are:

- Understand that past events often dictate current situations and dynamics.
- Identify the players in the game.
- Learn the rules of the game.
- Determine your own ethical bottom line.
- Develop a framework for making ethical decisions
- Understand that the political process can be as important as the substance of the issue.
- Focus on the issue, not on the personalities.
- Don't take the rejection of your idea or concept personally.
- Perseverance is a must.
- The most important thing to remember is: *fire chiefs don't vote.*

CHAPTER 7

Leading in Times
of Rapid Change

Over the past thirty years, the fire service has gone through significant change. As the industry strives to meet new objectives such as expanded missions, more extensive education, or greater safety, our environment has demanded change. At times, our reluctance to make these changes is a direct result of our failure to understand how we perceive, understand, and interpret a situation. I have a good friend who lives in southern California, an author and motivational speaker named Tom Bay. Tom became interested in the fire service several years ago and can often be found riding with the Orange County Fire Authority. In several of his books, Tom has written that each of us has a "life window" that we write on, which helps to provide a frame of reference on how we process and accept information. Our life window, therefore, dictates the directions that we as individuals take and will subsequently help determine where we lead our organization. Think about yourself and your own life window. What would be written on that window? What life events have formed your values, work ethics, likes, and dislikes? These all have a direct impact on how you process information, how you look at the world around you, and how you ultimately feel about an issue. During the past thirty years, innovation and improved science have required our industry to take a hard look at our traditions and our old ways of thinking and, in many cases, have challenged our individual frameworks. Change can never occur without a conscious effort on our part and a great deal of leadership, leadership that forces us to examine our frameworks and test them against the realities of the day. To do so, we must be willing to listen to others and be receptive to their perceptions.

There is a great story that demonstrates this on a very basic level. Two battleships assigned to a training squadron were at sea on maneuvers in heavy weather for several days. Sailors serving on the lead battleship were assigned to watch (lookout) duty on the bridge as night fell. Visibility was poor, with patchy fog and driving rain. Due to the bad weather the captain had remained on the bridge, keeping an eye on all activities. Shortly after dark, the lookout on the boom of the bridge reported a light bearing off the starboard bow. "Is it steady or moving extremely?" the captain called out. The lookout replied "Steady, captain." This meant that the ship was on a dangerous collision course with the other vessel. The captain then called to the signalman, "Signal that ship that we're on a collision course and advise *them* to change course by twenty degrees." Back came the reply: "Advisable that you change course immediately twenty degrees, Seaman Second Class Jones." By this time the captain was quite angry. "Would you tell Seaman Second Class Jones that his ship had better change course twenty degrees." He shouted to the signalman to also send the message "I'm Captain Smith, and we are a battleship. Change your course twenty degrees." Back came the flashing light telling the captain, "I am a lighthouse, Seaman Jones." The battleship changed course. Never let your rank or your ego take you someplace where you should not have gone. Each of us as leaders of our own ships and many of us as leaders in our industry are accountable for the welfare of many others. This alone requires us to be aware of, control, and take responsibility for some basic frameworks that we all know are part of the larger picture, as well as provide the leadership motivating others to adopt a more open and hopefully objective view.

Don't Play a New Game by the Old Rules

Albert Einstein once observed that the problems we face cannot be solved with the same level of thinking that created them. We all must realize the significance of the change that the fire service has experienced. We're in a different business today. With reengineering, consolidations, political shifts, lack of resources, increased service-level demands, and outsourcing of our services, the men and women who work in the fire service find themselves in perpetual change. It is important to remember that "In times of rapid change, experience can be your worst enemy." The rules of the game of public safety have changed, as has the whole of government. Whatever their position in the organization, leaders need to manage their own motivations to ensure that they can make a difference for teams, their organizations, and the community they serve. The steady waters of the past have given way to the turbulence of an open sea. It is a sea of change driven by performance and quality issues, where course corrections are the norm and can and must occur very quickly. As leaders, we have been challenged to help produce organizations that work better, cost less, and meet the safety and service needs of our local communities. After September 11, 2001, our mission expanded. We are the first responders, an integral part of the front line of America's homeland security and defense.

Today's leadership must not anticipate a future that resembles the past. We must focus on creating new opportunities rather than just solving the problems of today. Today's leaders must choose their own direction rather than doing the tried and safe. We must continually aim for something that will make a difference rather than something that is safe and easy to do. That is what leadership is all about. One of the problems of the fire service is that our heritage and traditions can often cloud our vision of what the future needs to be. In many organizations and with many chief officers, it is often difficult to step outside of the established framework and provide needed leadership for the future. This may be one of the most significant challenges for leadership in the fire service today: stepping outside of a framework that has developed over time, that integrates our values, our ethics, and our culture, and has a direct impact on our desires or dreams, both personally and professionally. In many organizations, this framework has become a driving force, dictating how information is perceived and acted on. Organizational norms become the filtration system, determining how information is shared, perceived, reacted to, and processed.

I've been in the fire service for over twenty-five years and have found that dealing with the organizational culture is the aspect of leadership that is most challenging, can be most frustrating, and can pay the most rewards. Breaking down the cultural barriers that exist within many organizations to provide a new level of thinking and openness is a mark of true leadership. Joel Barker, a futurist, has written several books on how paradigms (frameworks) affect our ability to lead change and the ability of people to accept it. Never before in the history of the fire service have we faced so many challenges as we do today. Never before have there been so many new service-level demands, technological choices, and new workforce challenges. Never before have we had the opportunities to make revolutionary shifts in our industry. Exciting, isn't it?!

It's all Up to You

How does one lead others to places where they would not have gone themselves? I often remind my staff officers that we're only as good as our last performance. There is an interesting comparison that I think sums up the world that we, as leaders, operate in today. Think about our competition as if it's the Super Bowl. We want to play hard for the season and win the big game, and then sit around during the off season and gloat about how great we are. We know that our competitors in business and in the public sector, won't wait until next year for the rematch. They want to play again, next week and every week, until finally they win. One of the leadership challenges, in both the public and private sector, is that the rules of the game are continually changing, driven by changing attitudes and values, technology, economics, community/consumer expectations, just to name a few. These also drive our competitive challenges. Within the fire service, these competitive challenges can be seen in the shrinking tax dollar and in the significantly increased demand for service, not only in relation to some of our basic tenets of EMS, Hazmat mitigation, and fire prevention, but also now, with the mandate to provide the first response to terrorist events. Our competitive challenge is seen in our governing bodies, which continue to expand the scope of services we provide, with a limited or no increase in the resources available at the local, state, and national levels.

> Through all of these changes, as chief officers we have the opportunity to be evaluated and judged on how we dealt with each one of those issues. We can have a 20-1 record, but our leadership may often be judged by our latest performance.

I've seen fire service leaders who have unified organizations in disarray. I've also seen leaders take a good department and make it great. To be successful, they realize that organizations need to be refueled, revitalized, and refocused periodically. Leaders understand the need to periodically create new visions as a vital link to developing high-performance, high-quality, service-oriented teams, leading their departments to what is often referred to as the "cutting edge."

I have also witnessed several fire service organizations enter periods of stagnation in which they do the same thing over and over. There is very little creativity and innovation, and as time passes, there is the sense that they are lethargic in their approach and actions on many levels. The common denominator in each case is a fire chief who had a long tenure and whose leadership ability peaked several years earlier. Organizations that are not being led will settle into a comfort zone of mediocre performance. Where there is no vision and no motivation from the top to push the agenda, the organization will find a midline of performance.

This topic is well covered in a fine book called *It's Your Ship* by Captain D. Michael Abrashoff, former commander of the *USS Benfold*. It provides an insightful look at some basic concepts of leadership, motivation, and productivity. It's the story of how Captain Abrashoff transformed the *Benfold*, a guided missile destroyer, in June 1997 from a ship that was mediocre at best to the best ship in the Navy as graded by its peers. Captain Abrashoff's book made it evident to me that there are parallels between what he faced when taking over the *Benfold* and what we face when we take over a new organization. He emphasizes the need to:

- Lead by example.
- Listen aggressively.
- Communicate purpose and meaning.
- Create a climate of trust.
- Look for results.
- Don't look for salutes.
- Take calculated risks.
- Go beyond standard procedures.
- Build up your people's confidence.
- Generate unity.
- Improve your personnel's quality of life as much as possible.

These are some of the same issues that we face as fire chiefs in our own organizations. These parallels are striking from a military standpoint, as well as from the standpoint of some of the bureaucracies that we face in making change happen and some of the politics within each organization that creates barriers to performance and success. Abrashoff offers an enlightening and original point of view on leadership, motivation, and productivity. He provides a commonsense approach to leadership and a fresh outlook on how to build an effective team. This is also an insightful story for fire chiefs who find themselves in the position of just "minding the store." The role of leadership is as important the day you come into the organization as it is a decade later. That is why it is so important to understand that if your performance drops, so will that of your team. It is up to you to keep yourself on the leading edge through continued education, fire chief association involvement, teaching, mentoring, or whatever keeps you at the top of your game.

Mayberry RFD

We have focused considerable attention on individual teams within the organization, but what of the organizational team as a whole? I noted earlier that in the absence of leadership, organizations will arrive at a midline level of performance, which is driven not by the pursuit of excellence but by the preservation of the status quo. Another problem is that the entire team becomes complacent in their attitudes and actions.

This gradual decline of leadership is a direct result of losing touch with the organization, resisting innovation and change, and failing to take advantage of technological breakthroughs. If you look at a fire department as a system, there are many forces and interrelationships that impact it. These are presented in the following diagram.

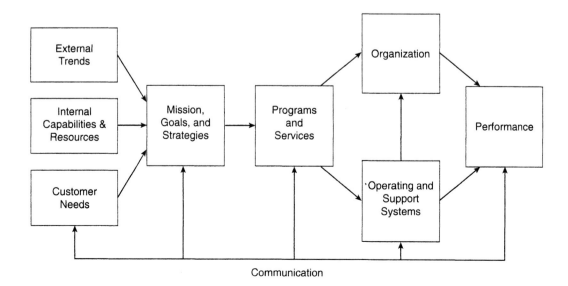

In a department like this, change is often triggered by external events. It could be a significant fire, lawsuit, political election, or new city management. I've known many fire chiefs who have lost their positions because they failed to recognize the need to refocus their organization. Complacency left them vulnerable and created organizational apathy. When organizations are out of touch with the changing context of the industry and the community they serve, they find themselves in an organizational comfort zone. It's an especially vulnerable place that experienced chiefs must avoid. It's a place where nothing much happens, where the need for bureaucracy, fellowship, or socialization is greater than the need for innovation. The need for progress and new ideas in planning isn't recognized. It's what we call the "Mayberry syndrome" of the fire service. It's a nice place to visit where very little changes. But it also is a dangerous place for our profession and the firefighters who work there.

Such complacency often leads organizations into what has been called the "boiled frog syndrome." This describes what organizations experience when they become complacent about what is occurring within and around them. If you take a frog and place it in a pan of boiling water, it will immediately jump out. More often than not, it will survive the experience. If you put that same frog in a pan of room-temperature water and then heat the water very slowly, the frog will stay in the pan until it boils to death. The frog could have jumped out at any time, but the change happened so slowly that it didn't realize the danger. Our organizations can react at times in the same way, becoming complacent to the point of self-destruction. Those that don't respond to internal or external forces in time to avoid significant damage often find themselves in the same situation as the frog in that pan of water. These are organizations that are "playing fire department," and then something catastrophic happens as a result of their complacent attitude. In today's fire service, there is no room for such organizations. With approximately 30,000 fire agencies across the United States alone, the range of ability can vary greatly. Unfortunately, the expectation of the community—whether it is a small rural community in the middle of Indiana or a major one like Chicago— is that the organization is fully equipped, well trained, and ready to handle any emergency. We, as a fire service, have fooled ourselves into thinking that we can actually provide that level of service in all of the communities we serve. In reality, that is not the case.

A report that the National Fire Protection Association published in 2003 provided an overview, a look at the needs of the industry, and an assessment of the fire service. This work was funded by a noncompetitive grant of the Firefighter Assistance Act, also known as the Fire Act. This legislation also provides grant money directly to fire agencies for equipment, training, and related fire prevention programs. The program was created in 2001 after Congress passed legislation initially allocating $100 million for direct grants to fire departments. Grant applications are reviewed by panelists from the fire service organizations, based on the financial needs of the applicant and an analysis of the benefits that would result from the award. Another aspect of the Fire Act is research, and this report indicates a number of organizations that are performing services they are not trained to provide. This is a good example of the Mayberry syndrome. For example, as reported, over 99 percent of U.S. fire agencies stated that, as part of their mission, they are structural fire attack agencies. Yet 90 percent are not trained to do this. The same is true of other areas such as EMS, confined space rescue, specialized rescue, and Hazmat handling. As chief fire officers we must ask ourselves, how widespread is this organizational complacency within the fire service today?

I have witnessed many fire chiefs lose their position due to organizational complacency. In most cases, I have observed, this happens because of a significant event in the community. It's directly attributed to the lack of preparation and preparedness by the fire agency, which directly reflects back on the leadership of the chief fire officer. In many of these cases, a significant fire occurred, and because of lack of training, lack of fire ground discipline, and lack of needed equipment, the community suffered a significant fire loss, a significant civilian fire death, or, in many cases, a firefighter fatality. In all of these situations, it was the chief fire officer who had to defend his or her actions, the actions of the community, and the actions of the firefighters. In many cases, it was due to their complacency as leaders—not challenging the status quo, not challenging the cultural aspects of their organization, not challenging city and county leaders to provide sufficient funding to ensure that they could do their job correctly. This complacency may be one of the biggest risks that a fire agency faces today. Continued organizational revitalization is essential. It is the hallmark of successful leadership.

Kent Leydan, an organizational development specialist from Oregon, studied what groups go through as they process change. He observed that there are five phases in group dynamics—forming, norming, storming, performing, and mourning. As we look at our organization from a team perspective, we must have the ability to persuade our employees to take the risks and make the sacrifices necessity to achieve the goals of the organization. Sun Tzu's book *The Art of War* is not really about war, but rather a practical guide on how to plan and how to use your personnel effectively. Tzu writes, "the way means inducing your people to have the same aim as their leadership so they will share death and share life without fear of danger." In short, followers must have the same convictions as their leaders to be effective and get the most from the team. We must understand the dynamics that our entire organization and specific work groups experience, sometimes on a daily basis.

I know from personal observation and experience that as you institute change within the organization, different groups will react differently. Understanding that as a chief fire officer is very helpful in reducing the tensions that sometimes accompany the change process and helping the group focus on achieving a higher level of performance.

Groups

Groups, like organizations, are dynamic. Each group will continually evolve and assume different behaviors and positions, depending upon what is occurring within the group, the personalities involved, and the issues at hand.

Groups

How and Why Formed
- Enable tasks to be done faster and better.
- Function only if members receive interpersonal support.
- Must attend to members' social needs as well as to tasks.
- Must maintain some degree of cohesion.

Task Behaviors
- *Initiating:* Giving ideas or directions, suggesting goals, defining problems, suggesting ways to solve problems.
- *Clarifying:* Building on others' ideas, interpreting, paraphrasing.
- *Coordinating:* Putting ideas together, organizing, developing an overall plan.
- *Summarizing:* Putting ideas together, providing focus.
- *Recording:* Keeping track of the group's activity and making the group's ideas public.
- *Evaluating:* Making critical judgments of ideas and suggestions.

Maintenance Behaviors
- *Encouraging and supporting*
- *Harmonizing:* Mediating differences, relieving tension, working through problems.
- *Gatekeeping:* Keeping communication open, making sure everyone has an opportunity to participate.
- *Observing:* Keeping track of how the group is functioning, providing feedback to help the group function more smoothly.
- *Following:* Accepting others' leadership, indicating support and agreement.
- *Setting standards:* Defining norms to which group members adhere.

Phases
- *Forming:* Individuals begin to see themselves as a group with a common task. Issues are involvement and cohesion.
- *Storming:* Group members vie for leadership and control.
- *Norming:* The group develops expectations about acceptable behaviors and an identity that sets it apart from other groups. Issues include agreeing on group expectations and how to deal with members who do not meet them.
- *Performing:* Group members work together actively to solve a problem or perform a task. Issues are individual performance and coordination.
- *Mourning:* Group members share their common past (debrief) and prepare to go their own way.

A noted author of the 1980s and 1990s on organizational leadership and change is Tom Peters. His book on organizational excellence, coauthored with Robert H. Waterman in 1982, shows how the window on the future lies in the occurrences of the past. As I read his book, *In Search of Excellence*, I was disturbed to find that many of the corporate leaders of that time and many of the organizations that were experiencing so much success when Peters wrote his first book, leaders noted for their vision, had either been replaced, or their companies had suffered significant losses or had gone out of business in the 1990s. The message is worth noting. To remain successful, leaders have to understand that change is a continual process. Transformational leaders often find themselves leading the charge in an attempt to get people and organizations to expand their frameworks and change old methods of doing business before it is too late.

The organizational shifts of corporate America noted in Tom Peters' two books probably also occur in many fire service organizations. I think we would find that many organizations are in a state of complacency or disruption due to a lack of leadership and/or inability to accept and deal with change and innovation.

Leaders who recognize the need to revitalize their organizations must create a new vision for them. In many cases, this means expanding the frameworks of the personnel to allow new information, thoughts, and processes to work. As fire chief, you have to develop a vision that is acceptable, workable, and assumes a desire for change in your organization. Of equal importance, you must be able to communicate this vision in a way that matches your own philosophy and leadership style. We have to recognize that despite the significant changes of the fire profession of the past thirty years, "we ain't seen nothing yet." I predict that in the next twenty-five years the fire service will be called not the "fire service" but the "department of emergency services". In fact, we could change our name today, because each year the percentage of fire suppression activities related to the other functions occurring within the fire service is much less than it was a decade ago and will be even less a decade from now. In 2020, the issues facing local fire chiefs, and the leadership at the national and international levels, will be drastically different.

Of equal importance is how we will continue to professionalize our industry, not only at the chief fire officer level but throughout our organizations. The fact is that we must continue to raise the bar if we are to lead the changes in fire protection and emergency services in the future. We must not become good at what is worse for our organizations: the failure to look ahead and begin to revitalize and revamp our organizations to meet the challenges and opportunities of the future. As leaders, we need to perceive the role of fire chief as involving much more than dealing with the day-to-day tasks of running a department. What is needed now and in the future are transformational leaders who can define what the issues will be. We must be aware of that responsibility if we are to take on those issues. As we accept the need to move our organizations to a higher level of performance, we know that it will require personal risk. Leadership is never risk-free. On the other hand, neither is complacency. As we prepare our organizations and people for what is to come, we have to understand that as we try to expand our organizational vision, we often find ourselves the only person on the playing field—at least initially.

It is one thing to have a vision but quite another to motivate people to buy into it, believe in it, follow us, and ultimately promote it. As leaders, we must recognize that the status quo is never enough. In doing our job, we must push the limits of the comfort zone of the organization and the people within it. Be aware that as change occurs, people in the organization will try to reestablish their comfort zones, often through unique and in some cases bizarre behavior—grievances, starting rumors/gossip, and passive-aggressive behavior toward you, individual programs, and the organization. In developing a new framework and a vision for our organizations, it will be critical to maintain a common thread: the value of the organization, the ethics that we operate by, the mission, and our long-term objectives must be clearly articulated. Warren Bennis, in *Why Leaders Can't Lead* (1990) and *On Becoming a Leader* (2003), outlines the differences between a leader and a manager. There are several differences, and they are crucial as we discuss group dynamics and leading change. As you read the following list, ask yourself which one you are. If you have the title of fire chief and find yourself to be a manager rather than a leader, you need to step back and retune your own method of operation. As Bennis puts it, *leaders conquer, while managers surrender*. Look at the differences:

- The manager administers; the leader innovates.
- The manager is a copy; the leader is an original.
- The manager maintains; the leader develops.
- The manager focuses on system and structure; the leader focuses on people.
- The manager relies on control; the leader inspires trust.
- The manager has a short-range view; the leader has a long-range perspective.
- The manager asks how and when; the leader asks what and why.
- The manager has his eye on the bottom line; the leader has his eye on the horizon.
- The manager accepts the status quo; the leader challenges.
- The manager is a classic good soldier; the leader is his own person.
- The manager does things right; the leader does the right thing.

Leaders have a clear idea of what they want to do both personally and professionally. That strength and tenacity should be persistent in the face of setbacks and failure.

Stewardship and the Core Value of Service

Leadership is more than command and control and, in the fire service, having command presence. You can have those attributes, but you may not be effective, appreciated, or followed as a leader. I have found that the leaders who ensure that service becomes a core value of the organization are the most successful. Leaders who do this will lead by example and will institutionalize the basic concept of service over self. I've found that organizations in the fire service and in the private sectors that embody this core value of service in their day-to-day operations are truly the best organizations. Think about it. Would you rather buy a product from a company whose core value is to make you happy or one whose core value is to make a profit, no matter what? Service-oriented organizations also have a broader perspective on what is going on around them. They are not self-absorbed, they understand community dynamics, they have the flexibility and ability to change rapidly, and they have some of the brightest people because they attract the best applicants.

In the fire service, we need to make sure that our basic skills, our core competencies, are relevant. These core competencies focus on what has to be done from a technical standpoint, but the core value of service goes beyond our ability to respond to an emergency once the alarm goes off in the firehouse. The core value of service embodies who we are and the expectations we have for one another. Our acceptance that stewardship is a core value underlying the services we provide is the basis of our ability as an industry to grow. Keshavan Nair counsels Fortune 500 companies on leadership and decision making. He claims that the core value is not just a policy but something deeper and more profound. It states that every individual in the organization values being of service. When service to others is valued, trust, loyalty, and truth flourish. These create efficiencies of trust.

I believe service, leadership, and stewardship are intertwined. In watching successful fire chiefs, I have found that those who have this core value of stewardship are the ones who place service above self and truly believe that organizational service is a core value—a self-sustaining value and one that has tremendous power. Keshavan Nair also writes that the acceptance of a core value implies a commitment. Core values within an organization and core values in service are long-term commitments. When you have both belief and commitment, you and the organization can achieve great things.

Core values often become the foundations of the traditions that will exist in the future. Once you as a leader have articulated that service is a core value, the challenge is to determine how you can begin to influence others in your organization to act in a way that supports that core value. It starts with you, leading by example to ensure that you are doing more than you are paid to do. The model of good leadership is the most effective way to inspire others' commitment.

Albert Sweitzer said "Example is not the main thing in influencing others, it's the only thing." That is certainly true in this case.

In promoting stewardship and service as a core value, there has to be a commitment to and from everyone in the organization. We need to recognize those persons who provide an example of good stewardship, of service over self. I have used a document called the "Hot Sheet"—a weekly two- to three-page rundown on what is occurring during the week within the organization. This document is e-mailed and faxed to all facilities, as well as to persons in government in the nearby cities and counties. It's an excellent way to showcase people who are models of the commitment to service and helps to instill this core value of service. Here are a few examples of how this works. An engine company responded to a call from a man who fell off a ladder while fixing his gutter. As a result of the fall, he had broken off a piece of his deck and had also broken a couple of ribs. After the crew arrived and took care of the man, they stayed a little longer to fix his deck. When the man came home from the hospital, he found that it had been repaired.

Another story involves an eighty-year-old man who was raking leaves and experienced chest pains. When the crew arrived, after taking care of his medical situation and placing him in the ambulance, some of them stayed behind and continued to rake the leaves out of his yard. When he came home, he found that his yard work had been completed. Still another story concerns an elderly woman who was inappropriately burning yard debris in her back yard. The local fire crew was called out. They notified her of the fire code violation and extinguished the fire. The next morning when the captain went off duty, he stopped by with his pickup truck, cleaned up the debris, and removed it.

Creating this type of stewardship is an essential step in developing a core value of service, which I believe is essential to making change work for us in the future. It transcends the day-to-day issues that we face and helps to maintain the foundation on which the fire service has been built for the past 200 years. I was recently doing a promotional interview with a captain candidate whose name had repeatedly come up for providing good customer service and creating a stewardship process in his company. I asked him how he had created the idea of service over self. He indicated that it was a form of competition among his crew members to see what more they could do for the people in every emergency call they responded to. Sometimes it was as simple as putting up a free smoke detector; at other times, it meant cleaning up the kitchen before they left if the occupant was going to the hospital or ensuring that the house was locked and notifying the neighbors of the situation. In each case it was different, but it was clear that the concept of stewardship had become a core value of this company. It was reflected not only in the comments of the general public but also in the recognition of the ability of other crews to go above and beyond the call of duty. Vince Lombardi, the late coach of the Green Bay Packers and football coaching legend, said, "Excellence isn't a sometimes thing. You have to earn it and re-earn it every single day." Promoting excellence every day has to be our goal to give our customers what they expect. As leaders, we have to focus relentlessly on the importance of achieving excellence, making sure that everyone has it as part of his or her mission and is commited to attaining it.

Mark Twain may have stated it best: "Always do what is right. It will gratify most of the people and astonish the rest."

What We Can Learn from Others

In his book *Making the Corps,* Thomas E. Ricks provides insight into the successful transition of men and women who entered the Marine Corps. In just eleven short weeks, these young people were molded into focused individuals and in many cases instilled with the characteristics of effective leaders. What can we learn from them? Ricks found what he termed "lessons from Parris Island"—six key characteristics that have a great deal to do with leadership and offer powerful insights for anyone in a leadership role.

Of all the things that can motivate people, the pursuit of excellence is one of the most effective and one of the least used in our society. Yet the Marines, with their emphasis on honor, courage, and commitment, offer a powerful alternative to what we often see today. The lessons from Parris Island are straightforward:

- Tell the truth.
- Do your best, no matter how trivial the task.
- Choose the difficult right, not the easy wrong.
- Look out for the group before you look out for yourself.
- Don't whine or make excuses.
- Judge others by their actions, not by their race or gender.

These lessons provide a very simple bridge to leadership. Concentrate on doing a single task as simply as you can. Execute it flawlessly. Take care of your people and go home. These seem to be basic steps, yet they are powerful in running an organization. They demonstrate the organization's leadership, its core value of service, and its commitment to pursue excellence in all the things it does.

Paralleling *Making the Corps* is a book written in the early 1600s called *A Book of Five Rings.* The author, Miyamoto Musashi, was a teacher of the samurai way of life—a philosophy and an approach for those who desire to construct a full, dynamic, and successful life. It speaks to me in the same way as *Making the Corps* regarding what's important in a leader. The way of the samurai consists of the following principles:

- *Act:* There is no perfect moment when the action must be done; concentrate on the action.
- ***Don't complicate the situation:*** Emphasize simplicity, naturalness, and down-to-earthness.
- ***Focus on your purpose:*** What do you want to accomplish and what do you want to be? Find out what is important to you personally. Life is short, so confront it immediately. Whatever you do, think of what you are trying to accomplish.
- ***Always advance on life, never retreat from it:*** If you feel shy, overcome the shyness so that you can advance. Remember, it is better to die from a sword in the chest rather than an arrow in the back.

• **Leap into action:** If something is to be done, do it. Risk defeat. When you avoid action, you spend energy dealing with the unknown rather than on the job at hand.

• **Be absorbed in the action:** If you are absorbed in yourself, you tend not to act. You back off from the action, whether it involves making a decision, selling an idea, applying for a job, or making a career move.

• **Concentrate more on the action than on yourself:** It can be more effective. If you back off and think about what could go wrong or whether you made a bad decision, too much time will pass and the opportunity will be gone.

Making the Corps and *A Book of Five Rings* offer a historic parallel regarding leadership, value, and service. They each demonstrate what leaders in today's fire service need to focus on if they are to be successful in promoting organizational change by establishing a core value of service and stewardship.

What both books point to is the ability of leaders to create a dynamic within their organizations. It can best be summed up by the saying "We must leave it better than it was when we found it." What we are trying to create in all of our organizations, whether public or private, is "above-the-line accountability." It expresses the idea that "If you see it, you own it. If you own it, you can solve it. If you can solve it, you should just do it."

Above the Line—Steps to Accountability

The opposite of above-the-line accountability is the victim mentality that we often see in many of our organizations. Those who have it ignore or deny, say "It's not my job," finger-point, to protect themselves, not get involved, wait to see what happens, and create confusion through rumors and passive-aggressive behavior.

All the discussions that we have about leadership and group dynamics can be summed up by the concept of above-the-line accountability within your organization. Then excellence won't be a sometime thing; it will be an everyday thing. That's what we, as leaders, are trying to do every day. As leaders, we must never underestimate the power of our actions. Even the smallest gesture can change a person's life, for better or for worse. Our goal must be to create conditions that produce commitment and creative action by the people in our organizations. When we can create ownership using above-the-line accountability, we will be most effective. It's a simple and straightforward concept, and something that your leadership team can easily grasp. If you see it, you own it; solve the problem. If you do anything less, you're not acting as a leader. I've found this idea to be very effective and useful, and I've seen it become a core value for organizations I have worked in.

Agenda for Change

The past is gone: The present is full of confusion: and the future scares the hell out of me.

DAVID L. STEIN

Going through any major change will challenge the way we view ourselves. Big changes can be like death and rebirth for an organization. Living through this process is often like undertaking major remodeling in your home. To achieve the result you want, you must first rip out the old, leaving the basic structure, and begin to bring in the new. Once you've added the final touches, you can move back in, begin to feel comfortable, and once again become productive. Change is a lot like that, because it always takes a little longer and costs a little more than you originally thought. Structural and cultural changes require people to let go of the old ways and live through a period of doubt and uncertainty. Managing this process takes sensitivity since change can be frightening to employees. We all know that change is inevitable. Since this is the case, a good question is "Will we always be riding this wave of transition?" The answer is probably "yes." For without change, we and our organizations would become stale and unresponsive. The challenge is learning to move through the transition as easily and creatively as possible. What helps people navigate through this unknown territory is a map of what they can expect and information on how they can respond most effectively to challenges as they occur.

As fire chiefs today, we face a unique challenge. The old adage of the fire service, "Two hundred years of tradition, unhampered by change," is simply not true anymore, and I don't know if it ever was. Yet, one of the challenges we must face, from an organizational standpoint, is to get our personnel to look past the four walls of the fire station and understand that the world around us is changing and placing significantly new demands on everyone within the organization. As fire chief, you have to understand the resistance to change; it is part of our culture, and for most people it is their natural tendency. Change is easier said than done, and whether we try to alter our environment, our workplace, or the services we offer, the change destabilizes our organizations. In some cases, it creates fear. A sense of permanence and tradition has been the backbone of the fire service and has become a significant influence on the culture of the industry. This permanence and tradition have provided stability for our organizations, but in today's changing environment they have also produced many organizations that are inflexible, unwilling to change, and bureaucratic in their approach. It's safe to say that the pace of change will only increase, and as we move our organizations into the future, we must have our engines running at full throttle and a firm hand upon the wheel.

Faith Popcorn, an author on change and trends, has predicted that we will change as much in the next ten years as we have in the past fifty. Consider the ramifications. She states, "From 1940–1990 we experienced fifty years of change. From 1990–2000 we experienced fifty more. From 2000–2005 we'll experience fifty more." What happens after that? If you plan to be part of the fire service for another fifteen to twenty years, you may live through exponential change unlike anything we've ever seen or experienced before.

The need for change is evident. Just look around. Almost everything in our environment is changing; therefore, we must change with it. Remember the old saying, "If you always do what you've always done, you'll always get what you've always gotten," and in some cases, maybe less. This holds true for all of us in the fire service and for most organizations today. To successfully navigate the changes that will be needed under your leadership, you must understand the importance of culture, vision, and the need to possess a global view. The ideals that we must promote to be successful—the importance of our leadership, empowerment of others, the ability to measure our performance, and understanding how change affects our people and our organizations—are critical elements for you as fire chief.

Nature of Culture

During times of great change, we have to ask our employees to be flexible; if we don't, we cannot position ourselves today to be competitive in the future. Forces of change that we'll experience in the fire service will come from both inside and outside the organization, with demands from our constituents and changes in the political arena that we work in. As we move through such changes, it's crucial to understand our organizations, their culture, and their effect on the change process. Whether change is initiated from inside or outside an organization, culture plays a large part in determining how it will be processed, accepted, and ultimately dealt with. We find that cultures and strategies often change, or conflict with one another, when change occurs. When conflict occurs internally, when the organizational culture doesn't embrace the changes you've initiated, the change efforts will often involve a struggle and will not be successful.

In contrast, when the change is externally driven, both the organization and the fire chief can find themselves in extremely vulnerable positions. In many cases, this form of change is often the most dangerous. Whether it's political, economic, or service driven, people find themselves in a crossfire where no one wins. Often I've seen change caused by consolidation, or a need to save money or to expand service, met by organizational resistance. In each case, the change effort did not fail, but the organization and its culture were permanently altered.

Such a change was occurring as I was writing this book. For the past thirty years, the fire service has been expanding its role in emergency medicine— from first responder to basic life support (BLS) to advanced life support (ALS) to performing vaccinations and other community health care services. Today, EMS accounts for the majority of calls that we respond to. Yet, many metropolitan-sized organizations have resisted the incorporation of EMS into their fire service deployment package. In many cases, the resistance was due to the culture of the organization, which was so ingrained that even though the number of fires continued to decrease, the culture had created a level of arrogance that did not allow these people to see the changing context of the industry and their own environment. Then the terrorist attacks of September 11, 2001, occurred, and after twenty-four months of economic downturn, many cities and counties were facing significant budget shortfalls. Many of these fire agencies experienced reduced staffing and related cuts, including firefighter layoffs, to balance the budget deficits.

Here is the interesting fact: those metropolitan departments that were targeted for cuts had a common denominator: they did not provide ALS service.

So, it came down to a value-added question for the local governing body. This is a good example of how the culture can move an organization or keep it in a place where it should not be. In larger, older, more traditional departments, it is often difficult to make a shift in services due to the culture of the system. Throughout our industry, it is evident that EMS is now and will be a critical element of our service delivery package in the future. Why did those organizations not see that? The answer is that they saw it but could not move the organization in the new direction. Why? Culture! Guess what? They are moving in that direction now!

In many of these situations, it takes years to overcome the emotional upheaval that change creates within the organization. The future fire service must promote a culture that is much more responsive, adaptive, and accepting of change because that is our future. It's important to understand the responses to change. Organizations, culture, and people respond to the pressure to change in vastly different ways, ranging from enthusiasm to indifference to fear to anger.

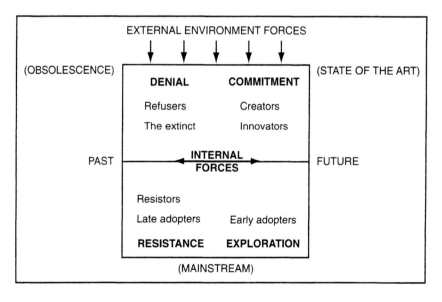

People who have a firm commitment to the organization and are motivated to stay on top professionally are usually on a continual quest to be at the cutting edge in everything they do. These are the "innovators," the most creative people in the organization, the first to embrace new ideas. Their commitment to achieve a competitive edge through knowledge is very high. I often refer to these people as "lead ducks"; they are often shot at, but they are usually the first to arrive.

In the next category are those I call "explorers" or "early adopters." These are the people in our organizations who have a high degree of curiosity and a strong sense of mission. They are also on the cutting edge of change and don't mind taking a few risks along the way. They tend not to be lead ducks, but they are definitely near the front of the formation. The next category consists of the "late adopters." These are the people who are somewhat resistant to change; they are slightly timid, with a desire never to be wrong. They wait for ideas to be well proven before they jump on the bandwagon. They like to be in the safe zone, and their objective is to reduce their vulnerability and risk.

The next group is the "resistors." Those people are uninformed and want to stay that way. They work hard to maintain the status quo based on their apathy, negativity, or simple ignorance. Change can overcome the resistors, but it often takes an enormous amount of organizational energy to do so.

The last category, where we often find far too many people and organizations in the fire service, consists of the "refusers." They have chosen not to progress. They refuse to accept any change, whether it's within the organization or in the world around them. That's why we often refer to them as people who are or are about to become extinct. They choose to follow the road of the past that leads to a community called obsolescence. They live in a time warp, and the only way to deal with their denial of the change that is occurring is to wait for or motivate them to retire.

Whether an agenda for change is motivated by internal or external forces, there are some basic requirements that you, as a leader, will need to address in order to succeed:

1. Top management must be involved. They need to set the example and be active in the change process so that others in the organization recognize their commitment.
2. Measurement systems must be used to track the progress of the change at both the upper level of the organization and in day-to-day operations.
3. You, as the leader, need to set the bar high and push your organization.
4. You must understand the need to provide education on how and why the change has to occur and the route you plan to take.
5. If you've implemented a change and it's been successful in your organization, spread the story within your organization and to the entire fire service.

Leading a cultural or organizational change involves four key issues:

1. *Information*—what is the change?
2. *Inspiration*—why is it needed?
3. *Implementation*—how will it be done, both individually and organizationally?
4. *Institutionalization*—how do we know when we've succeeded?

If you deal with these four aspects of cultural change, you will be able to start leading your organization in the direction you want to go.

Vision Is a Must

Give to us a clear vision that we may know where to stand and what to stand for. Let us not be content to wait and see what will happen but give us the determination to make the right things happen.

PETER MARSHALL

When we talk about organizational change, one of the best places to start is with vision. That has to begin with you. You can't impose your vision on others, but your clear view of where the organization needs to go can be a road map for many and an inspiration as well. A good place to start is by formulating a vision statement—a declaration of the organization's most desirable future. In the fire service, most of our mission statements are very similar, but our visions for our organizations in five to ten years can be vastly different. You as a leader must make the vision part of your game plan. Strategies that develop from a vision of how the transformation will occur are extremely important. A vision statement must also be emotionally charged. As a change agent, you must figure out a way to motivate your personnel to deal with that long-term goal as a necessity for the organization.

As president of the IAFC in 2002–2003, I had the opportunity to transform my vision into an action plan on a national basis. I chose the theme "A Time to Lead," as I believed it captured the new sense of purpose, pride, motivation, and challenge that the fire service faced after September 11. It also expressed what the leadership of the fire service needed to do at this moment in time. The challenges that we were facing with the expanded mission of homeland security, our need to clearly articulate our needs on Capitol Hill, and the economic downturn of the economy required leadership. "A Time to Lead" became the theme of the IAFC with a focus on a common vision. What was fascinating to observe over the course of that year was the number of other organizations that began to reiterate and make this vision their own. From IAFC divisions to committees to state fire chiefs' associations, the theme was repeated over and over. The result was more involvement of individual fire chiefs in the political process, more focus on common issues, and more recognition on Capitol Hill of the IAFC and the issues of the fire service.

The fire service of the future must remain competitive with the private sector, maintaining the adopted level of service within all jurisdictions, and meeting the goals, demands, and expanded missions of our customers. A clear, concise vision of where the organization needs to go is essential to making that happen.

Making the Vision a Reality

As the chief fire officer, it is never easy to turn your vision into action. As we implement change, success will depend on everyone's understanding both the need for change and the change process itself. It's important that we match human issues with organizational issues. How you lead change is often dictated by the culture of the organization. If an organization readily accepts radical change, multiple changes can easily happen all at once. Conversely, in an organization that has no experience with change or is resistant, changes must be instituted slowly and gradually, with strong efforts to build a consensus. A key aspect of the change process is understanding that some of the most difficult challenges involve the refusers. On the one hand, it is very important that we continue to keep an open mind and treat them fairly. On the other hand, you as a leader have limited time and energy. In my experience, your efforts will be better spent trying to win over the resistors and those who have already made a commitment or who are exploring various options. These are the people who will be much better served by your focus on them and on the issues at hand.

When we speak of change, we need leadership, not management. Change requires top-down commitment, not only from you but also from your senior staff, or the process simply won't work. There is a need to have buy-in from all levels of the organization and gain a consensus on a vision statement. This is a critical aspect of the change process. You can accomplish a great deal when everyone is geared toward achieving the vision defined within the organization. Developing a common focus that everyone can understand requires strong leadership that continues to articulate the vision. The ability to bridge the gaps between human and organizational issues is essential if we are to be successful in leading change. Change requires leading by example, flexibility toward the organizational structure, and an emphasis on team leadership. Change provides an opportunity to guide your organization, but it is not something that you can totally control. However, you can anticipate, adapt, react, and act accordingly so that your chances for success are enhanced. Chief fire executives today have to begin planting the seeds of change for the future, providing a means to move their organizations toward the change process and successfully implementing that change for the organization and its personnel.

Building a Culture of Our Own Choosing

There is an interesting analogy to how cultural norms can impact and play out in our organizations. It's the story of the five apes.

Start with a cage containing five apes.

In the cage, hang a banana on a string and put stairs under it. Before long, an ape will go to the stairs and start to climb toward the banana.

As soon as he touches the stairs, spray all the apes with cold water. After a while, another ape will make an attempt with the same results—all the apes are sprayed with cold water.

Turn off the cold water.

If, later, another ape tries to climb the stairs, the other apes will try to prevent it even though no water sprays them.

Now, remove one ape from the cage and replace him with a new one. The new ape sees the banana and wants to climb the stairs. To his horror, all of the other apes attack him.

After another attempt and another attack, he knows that if he tries to climb the stairs, he will be assaulted.

Next, remove another of the original five apes and replace it with a new one. The newcomer goes to the stairs and is attacked. The previous newcomer takes part in the punishment with enthusiasm.

Now, replace the third original ape with a new one. The new one makes it to the stairs and is attacked as well. Two of the four apes that beat him have no idea why they were not permitted to climb the stairs or why they are participating in the beating of the newest ape.

After the fourth and fifth original apes have been replaced, all the apes that were sprayed with cold water are now gone. Nevertheless, no ape ever again approaches the stairs. Why not? "Because that's the way it's always been around here."

"That is how organizational behavior is indoctrinated into social/corporate policy and a culture becomes entrenched."

Unknown author

Have you ever experienced similar behavior in any organization that you have worked in or volunteered for?

Many fire service organizations today embody behaviors and norms that go back for generations. These cultural norms, in many cases, are the most limiting factor in the ability to address the issues that the organization is faced with today. Creating the right culture, articulating your vision, focusing on the mission, and planting the seeds of the future are critical elements of successful change. In *The Adaptive Corporation*, Alvin Toffler stated that "the adaptive corporation needs a new kind of leadership. These managers of adaptation are equipped with a whole new set of nonlinear skills. Above all the adaptive manager, today, must be willing to think beyond the thinkable, to pre-conceptualize products, procedures, programs and purposes before crisis makes drastic changes inescapable. Warned of impending upheaval, most managers still pursue business as usual, yet business as usual is dangerous in an environment that has become, for all practical purposes, in constant change." Today in the fire service, many of us have placed ourselves in a very dangerous position through our complacency about the future. The need to reposition our organizations for future growth, and the competition that our organizations face for revenue in conjunction with the ever-expanding list of services we deliver, have created a new reality for our industry. "Business as usual" is equivalent to a slow and certain death for many of our organizations and for you as fire chief. As we cope with growing expectations and with the increasing demand for services, our visions will become a critical factor in the future health of our organizations, providing leverage and influence with both the individuals and the organizations we lead. We know that the pace of organizational change will continue to increase for many reasons. You've probably already seen change since you've entered the fire service.

Learning new strategies on how to lead change will therefore be critical for the successful fire chief of the future. Fire organizations, like most individuals, won't change until they must. Historically, the fire service has had an unimpressive track record in responding to change. For people to change, they must adapt in three ways—physically, intellectually, and emotionally. It all comes down to the question "What's in it for me?" As a change strategist and as the leader of your organization, you must understand the reluctance you will face—the fear that will be created, the rumors that will start, the misunderstandings that will result when you begin to move people outside of their comfort zone. There is no one solution to managing change; every initiative has its own challenges, and every organization is different. Social and cultural changes have forced fire service leaders to reexamine not only how they lead their organizations but also how they manage them on a day-to-day basis.

While there may not be a specific template for change, there are several common factors that must be addressed as you and your organization move through the process. Here is a strategy that is not revolutionary or unique. It includes components that have worked for many organizations, both private and public, and it emphasizes what is necessary for successful change:

- Establish your need.
- Build relationships.
- Understand the problem.
- Research the past and look to the future.
- Commit to a solution.
- Continually evaluate your progress.

Establish Your Need

Successfully revitalized organizations start by communicating their vision on change using meaningful language. Classic examples are Motorola's six-sigma philosophy, General Electric's Boundary-less Quality Banner, and Ford's "Quality Is Job #1" or "Like a Rock." Each statement transmitted a message. Each message was about change and was a public declaration that the organization needed to rethink its old ways of doing business.

Inspirational leadership can often energize an organization to begin to accept this change by making public declarations that establish a vision and provide a target for the workforce. The task is for you to turn that vision into organizational needs and manage those needs to improve everyday performance. Establishing a need greatly depends upon how you sell the change. Is the change a result of a problem or a result of an opportunity? No component is as important to the outcome of any change process, or change initiative within the organization, as your own attitude. It's critical that you start with a clear understanding of the organization's culture as well as the organizational paradigm, the framework used to process information related to the change, and how you're attempting to build a new culture.

Build Relationships

The foundation for any change process is people. Finish this sentence: "My department could accomplish X if it weren't for_____or _____." You could probably select one or more names and/or organizational roadblocks. I believe that most organizational problems are people problems. To be successful as you introduce change into the organization, you must build a support system and find organizational champions who can carry the banner when you're not there. In times of change, people skills are definitely more important than the technical skills we bring to the job. The need to build your team, lead it, and develop strong internal support during times of change can't be overstated.

As we saw in the story of the five apes, there are organizations that, because of traditional behaviors, autocratic leadership, union leadership that promotes mediocrity, and/or old-style politics, will attack anyone who differs from the norm. This is a disease that afflicts many organizations today, and the only cure lies in the workforce itself. The fact is that the workforce (usually one person at a time) determines when the old way of doing business is no longer productive. I saw this occur in an organization where, due to repressive labor leadership, the employees were reacting much like those apes in the cage. If they tried to do something new, they were chastised for trying to be a shining star—a classic example of promoting mediocrity so that no one looked good or bad. Organizationally, this is like playing in a basketball game with one hand tied behind your back. When this is the culture, you can only hope to maintain your existing level of performance; it will be almost impossible to move the organization to the next level. The cycle is broken when one of two things occur: when enough workers are willing to change or when a sufficient number of retirements, hirings, firings, or some other type of attrition occurs to allow a new cycle to begin. The same can be true of the leadership of the organization whose culture consists of being exclusive, centralizing power, and treating employees without respect. The result is an unmotivated workforce that says nothing, never steps out of line, and just tries to survive the workplace each day. In either scenario, the resulting culture is not conducive to change or to addressing the difficult issues facing our industry today.

You need to know how much support you have and who your organizational champions are or will be. If you wait to build your relationships until after you've developed your game plan, you'll probably fail. Remember that building relationship starts with listening. Take the time to listen to what you hear, to really observe what you see, to take into account what you are feeling, and to understand that you trust your intuition. As you gain more experience as a fire chief, you'll become more effective at doing this, trusting your instinct about the right direction in which to go.

Understand the Problem

In trying to understand what the issues are, be aware that what might seem to be a problem may often be only a symptom of the real issue. Try to understand the cause and scope of the problem. It's like looking at an iceberg. There are often many hidden dangers under the water and out of sight. Going beyond surface issues and exploring your findings, keeping in mind solutions as you continue your analysis, yet guarding against imposing your own favorite solution is important. Don't fall into the trap of identifying or reacting to only those issues that seem immediate, or important, or the most comfortable ones to address. Your purpose is not to develop a final solution but to gain an understanding of the problem from a global perspective. Don't become distracted from the purpose of your change strategy. Quick thinking and reacting can result in the implementation of quick fixes, but in many cases these don't solve the real problem. They turn out to be only temporary.

In your early exploration of the issue, you're trying to determine the root cause of the problem.

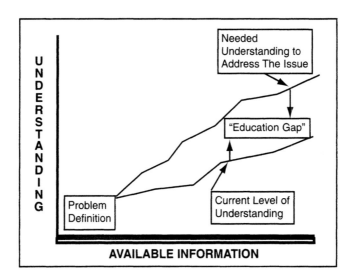

There's always a gap between what we know and what we should know before we set a course of action for our organization. To help address some of the cultural barriers and build the relationships that will be the foundation for your change process, articulating the problem and describing how it will be solved is critical. In every major organizational shift I have been involved in, there was an amazing number of perspectives on one topic: what the problem was and how it should be solved. That underlines the importance of understanding the problem. Once you understand the problem, you can close the educational gap within the organization between what you currently know and what you need to know to address the problem.

Research the Past and Look to the Future

None of us have all the information we need on a day-to-day basis to make quality decisions in an environment that is becoming faster-paced each day. It's becoming increasingly important to access information. The problem for many of us is that research itself is a science. Use both qualitative research, which draws on statistic surveys, questionnaires, and computer modeling, and quantitative research, which draws upon data from your direct observation. Don't dismiss your personal experience as unimportant. Look back first. Try to understand how your organization got to the point where it is today. Part of the reason may be tradition, so you have to know your organization's history. Consider the past solutions that succeeded and then review those that failed. Oliver Wendel Holmes once said, "To understand what is happening today or what will happen in the future, I look back." Today, most leaders who take control of the future do so by examining the past as well as by taking full advantage of the opportunities presented to them. Look outside the fire service;

many problems that exist within other types of organizations have similar characteristics. Evaluating and investigating other disciplines can be extremely beneficial. Networking in today's fire service is another excellent strategy. A big problem or a change opportunity is probably not unique to your organization. Somewhere another fire chief or chief executive officer has faced a similar situation and is an excellent resource. Reach out through fire service network to tap into this invaluable source of information.

Commit to a Solution

Colin Powell, in his book *My American Dream*, refers to a decision-making technique that relies heavily on intuition. He suggests that the key to not making quick decisions, but rather timing decisions, is based on a 40/70 formula. If he has less than 40 percent of the information needed to make the decision, he chooses not to act. Conversely, he doesn't wait until he has 100 percent of the information either, because in many cases that is simply not possible. He believes that when he has 60 to 70 percent of the necessary data he acts, trusting his intuition and experience. There are very few pieces of literature today for fire chiefs on strategies for implementing change and building a new culture. I think the basis for what we do in the future lies in the answer to the questions "Is it right for the organization, is it right for the community, and is it founded on providing good-quality service?" At times, our personnel can forget that and focus only on the issues that are central to them. As you put your research and findings into your own words and share them with the people you've identified as champions within your organization, stay focused on the task at hand. Every change initiative has its own challenges, and sometimes you have to write your own book as you go. As the fire chief, you have to be truly committed to the change initiative. If you're not, you'll likely fail. You have to lead by example much more than most people realize. Changing the organizational culture starts as a top-down activity, but it will be driven from below. If you observe companies in the private sector, those that succeeded in refocusing and reengineering made it because their chief executives were willing to invest their time and energy in the organization's change. Changing the organizational culture requires great courage and commitment from the chief executive.

Evaluate Your Progress

Evaluating change requires critical thinking and specific performance measures to be established. Without them, it is often difficult to evaluate your change initiative objectively. Organizational environments are in a constant state of change, and it's clear that yesterday's solutions won't meet tomorrow's challenges. Performance measures and the ability to quantify improvements made within the organization are critical not only to measuring the progress of the organization but also to articulating that you're on the right track. In his book *Taking Charge of Change*, Doug Smith outlines ten leadership principles and provides some insight into the change process. It's food for thought as you begin to move your organization into the future.

1. Keep performance results the primary objective of the change.
2. Continually increase the number of personnel taking responsibility for their own change (above-the-line accountability)
3. Ensure that people always know why their changed performance matters to the organization as a whole.
4. Put people in a position to learn by doing, and provide the information and support needed just in time to perform.
5. Embrace improvisation as the best path of both performance and change. Encourage off-the-cuff creativity and imagination, not rigid adherence to management plans.
6. Use team performance to drive change whenever required.
7. Focus the organizational structure on the work people do, not on the decision-making authority they have.
8. Create focused energy and meaningful language because these are the scarcest resources during periods of change.
9. Stimulate and sustain behavior-driven change by harmonizing initiatives throughout the organization.
10. Practice leadership based on the creed to live the change you wish to bring about.

Initiating a change process will take courage, vision, and a clear understanding of the culture of your organization. Sometimes it doesn't matter where you start, but as a leader your responsibility is to go first.

The Psychology of Change

Another crucial component of all change processes is the psychology of change. In his book *Managing Transitions, Making the Most of Change*, William Bridges provides a unique perspective on what people go through when coming to terms with change that is largely internal.

The transition starts by letting go of your old reality. Think back to some of the initiatives that you've tried to implement in your organization, and remember the difficulties that some of your personnel, and probably some of your senior staff, had grasping the new programs. The change can be a shift change, a new chief coming on board, or the addition of a new service. Some people are never able to make a successful transition. They simply can't accept the change because they've never let go of the old reality. When Bridges talks about the psychology of change, he describes three distinct phases, and as a fire chief I have seen all of them in the organizations that I have worked in.

The first is a period of transition that begins by letting go of the old reality. The second involves finding the neutral zone—the space between the old reality and the new one. This is the core of the transition process. It's the time when the old reality is gone but the new one doesn't feel comfortable yet. People who have made this transition are often in a state of confusion. This is when you can see the highest degree of mental anguish over having to let go of the old way of doing things. This is also when people feel most vulnerable. They are trying new things in a new way, and they feel this is the most dangerous time because things may not go as smoothly as they did in the past, at least for a while.

This leads to the last phase, which is actually the beginning of real change. It's the time in a person's or organization's life that involves developing new understandings and attitudes and often creating new values. It's the point at which you can begin to truly build a culture of your own choosing. New identities, new situations, new organizational values, and future leaders often emerge at this time.

How Do We Get Them to Let Go?

Making a change within the organization involves some type of transition. I've observed that in some cases, people don't resist the change; they resist the loss of something they've done for a long time, which removes them from their own comfort zone. Understanding the psychology of change is important so that we can identify early on what will be perceived to be lost. It's important to understand and describe in detail what the change will mean for all involved. Understand the chain reaction that will result when the change is made and who in the organization must go for the change to be successful. Two things are very important to understand: the current reality and the importance of the real and perceived loss to the personnel. In our leadership role, we tend to look at things from an objective and often a less emotional perspective. In many cases, our personnel may see them from a subjective and often an emotional standpoint. So, don't be surprised if, even on a minor issue, people seem to overreact because they are feeling a sense of loss. I've observed that people can overreact from an accumulation of change. Bridges refers to this as a "transition deficit," a time when the old loss has not been dealt with. All of a sudden, you see an explosion of emotion; it's the baggage of the past. The individual can carry only so much of it for so long until he or she gives out and lets go;

the result is an emotional outburst. Looking at the psychology of change, it's important that we acknowledge these losses openly, and have some understanding and empathy for the people who will ultimately have to make the change at all levels of the organization.

If you're making a large-scale change, you can expect to see classic signs of grieving within the organization. It's important to understand the components of this cycle. First is denial, second is anger, and third is bargaining or unrealistic attempts to avoid the change situation. Fourth is anxiety. Fifth is sadness and sixth is depression. Seventh is disorientation, confusion, and often forgetfulness. I think the best ways to overcome this natural process of grieving and loss is to give people as much information as you can, repeatedly if necessary. Define what has changed and what hasn't; if you don't, people will continue to use both the old and new methods. They will then decide on their own what they will discard and keep. The result will be chaos. Many people just toss out everything and wait for the organization to come back with a new set of changes. So, let your people take some of the old methods with them. It's important to be able to show what the end result will be so that you can ensure a degree of continuity as the change occurs. One of the most important reasons organizational change often fails is that we don't think of the end result. We do not plan how to manage the effects of change on our people. It's just as important to look for the emotional pitfalls as it is to understand the destination and how we plan to get there. A successful transition depends on our ability as leaders to convince people that it is safe to leave home.

The Neutral Zone

The neutral zone is a place between what has existed and what will be. There are some definite organizational dangers there. Anxiety within the organization is high, and the motivation level of many persons drops. There will also be some resentment toward you, as the leader of the change. People will become somewhat more protective of their territory, and their energy level will drop. But understand that this period will be short. It's also a time when people try to revisit problems that occurred as many as ten to twenty years before you arrived; change often reopens old wounds. Some people never forget, especially if they're focused on reliving the past or maintaining the status quo. When people are in the neutral zone, they feel that they are overloaded and getting mixed signals. The systems are in a constant state of flux. There's great confusion because they're having difficulty letting go of the past and coming to terms with the new reality. That often causes polarization between those who want to undertake the change and those who resist it. Polarization often places you in the role of counselor, referee, or, in some cases, autocrat, as you must make the difficult decisions and set people on the right course. When you lead change you'll find yourself in all three roles, sometimes daily.

It's also a time when the organization is trying to redefine and reorient itself. I have found that some things can facilitate this process. Make sure your mission is clear. Establish a network consisting not only of senior management personnel, but also of personnel throughout the organization to act as your messengers of change. Make sure you're accessible for people who want to talk. Don't forget to go to them. Don't skirt the difficult issues; meet them head on. Be honest and up front. Remain focused on what you're trying to accomplish because it's very easy to become distracted. Establish a positive outlook by your actions. This is the time to consider problems and propose creative solutions. Encourage openness within the organization. In building your team, this is an excellent way to start. Empower your people to experiment with new ideas. Turn your losses, your setbacks, and the obstacles that you've overcome into a positive message. Richard Nixon once said, "Only if you have been to the deepest valley can you ever know how magnificent it is to be on the highest mountain." When you're leading change, I can assure you that you will visit both places.

CHAPTER 8

Building Your Team

Robert Fulghum wrote *All I Really Need to Know I Learned in Kindergarten.* An excerpt from his book sums up what teamwork is all about.

> *Most of what I really need to know about how to live and what to do and how to be I learned in kindergarten. These are the little things I learned—share everything, play fair, don't hit people, put things back where you found them, clean up your own mess, don't take things that aren't yours, say you're sorry when you've hurt somebody, wash your hands before you eat, flush, warm cookies and cold milk are good for you, lead a balanced life, learn some, think some, draw, paint, sing and dance, play and work every day. Take a nap every afternoon. When you go out in the world watch for traffic, hold hands and stick together. Be aware of wonder. Remember the little seed in the plastic cup—the roots go down and the plant goes up. Nobody really knows how or why but we all like that. Goldfish, hamsters, white mice and even the little seed in the plastic cup—they all die and so do we.*

When I was growing up I remember the children's book *Dick and Jane* and the first word you learned, the biggest word of all, "Look." Everything you need to know is in there somewhere. It reviews the Golden Rule, love and basic sanitation, ecology, politics and sane living. Think of what a better world it would be if we all, the whole world, had cookies and milk at 3:00 every afternoon and then laid down with our blankets for a nap. Or what if we had a basic policy in our nation and in other nations to always put things back where we found them and to clean up our messes. It's still true no matter how old you are that when you go out in the world it is best to hold hands and stick together. Fulghum's commonsense approach is a poignant reminder of the importance of the basics in life.

Whose Side Are They On?

To be an effective leader, the fire chief must return to the basic premises that Fulghum writes about in his book. It's really about holding hands and sticking together. But getting to that point with senior staff and elected officials is often very difficult. Building your team can mean very different things, depending upon the culture of the organization, the level of trust, and whether you came from inside or outside the organization. It is amazing that on the fire ground we can demonstrate one of the most effective displays of teamwork known to any modern organization, yet those same people sitting around a table in a workgroup can be one of the most dysfunctional teams ever assembled. If you are to be a successful fire chief, building your team is essential. Assessing and changing the dynamics of existing groups within the organization is often not an easy task. I think you must work from the perspective of how you were selected. If you've been promoted from within, you have the advantage of knowing the organization's culture and history; the disadvantage is that you may have helped to create the very culture and history that, as the leader, you will have to overcome or possibly change. All the good things and all the bad things that happened to you in the past as a firefighter, engineer, lieutenant, captain, battalion chief, and now as the chief executive must be set aside as you deal with personnel and organizational issues. I've watched many chief officers whom I have worked with as peers and subordinates struggle to let go of what occurred in the past. The baggage that they carry in relation to certain personnel, to new ideas and innovation, and to change in general reduces their ability to be effective leaders. If you have come to the organization from the outside and you don't know its history, its people, or the dynamics of the past, it is prudent to be careful. Be wary of the messengers bearing insightful ideas during the first few weeks on the job. These ideas are often not presented with the best intent, and you can find yourself being misled, misinformed, and engaged in controversies that, as a new leader (or someone who is trying to build a team), you do not need.

I've given a lot of thought to the organizations I have led as fire chief. From a new chief executive's perspective, there are six areas that, I believe, are important to analyze in building an effective team.

1.   What is the level of trust in the organization? For most organizations, the level of trust is a window on the past. It's important for the new chief executive, whether coming from inside or outside the organization, to understand the level of trust and the reasons for it. Some organizations are so damaged by poor labor–management relations that a new leader, to be effective, will require a different approach. In this situation, a great deal of energy must be spent determining the basic psychological needs of your personnel and attacking the culture of the organization.

2.   How willing are your personnel to be honest with each other? It's a fascinating experience for a new fire chief to watch how the dynamics work, or don't work, in the department. This is especially true of the senior leadership. During the first three to six months on the job, a great deal of posturing normally occurs. During that period, you as the new chief are trying to determine the strengths and weaknesses of the organizations. Remember the SWOT analysis. What you'll find may surprise you. Many people simply can't tell the truth about where the organization as a whole stands or where their own division or department is in relation to the organization's missions and performance.

3.   Does your organization have the ability to communicate up, down, and sideways? I've found it very useful to watch how organizations, especially when I've come in from the outside, communicate at the upper level. This will tell you a lot about the personalities involved, their leadership style, and the organizational culture. An excellent organization has the ability to communicate in all directions. If there are communication issues, determine if they are symptomatic of the organization as a whole or if they concern one

person or one division. Organizations that allow communication upward are not only being led well but are also being driven from below. Organizations that allow only top-down communications are autocratic, hierarchical ("I have five bugles; therefore, we'll do it my way"), and myopic. Organizations that only communicate horizontally or sideways are being driven largely by their mid-level managers. Their leadership probably has no clue to what is occurring in the organization, nor does the firefighter riding on the truck. This creates a substantial void in the organization and is very disruptive. So, the nature of the organization's communication provides a great field of observation that, as a new or seasoned chief officer, you must continually pay attention to.

4. Does the organization kill the messenger? Trust in an organization means being able to be open and honest with each other. In some organizations, the bearer of bad news is the first one to go. This does not promote open communication, and reflects poor organizational culture and dynamics. If the bearer of bad news is punished, the messages will stop coming. This is a good measuring stick for your senior staff as you watch them for the first three to six months in your new position. You can tell a lot about these individuals by the way they handle adversity. If they have a tendency to kill the messenger, you should begin to work with them on that issue. If you do not, a serious situation may develop and you will be squarely in the middle.

5. Is the focus on problem solving or decision making? As you watch your senior staff in the first three to six months, you need to make an evaluation. Are they decisive or indecisive? How do they process information and solve problems, if at all? I've seen people who, once they pin a gold badge on their uniform, find it easy to make all the decisions. This is not a productive way to empower the organization and create a team. On the other hand, if you're the new chief, indecision is what you may find simply because people are trying to figure you out, just as you're trying to figure them out. Since they don't want to step over the line, they will wait until they can determine your leadership style and what you expect of them. When you begin to build your team, what you want to avoid is a workforce that doesn't speak up because they fear to be ostracized or cut off from the information flow. If the majority of the workforce is unwilling to speak to the people who can help solve problems, the leaders and managers of the organization have an inadequate pool of information for making quality decisions. This situation, in turn, often fuels workers' perception that managers are untrustworthy or incompetent. Therefore, it is extremely important that you and the top leaders, as a team, begin to consider how you process information to solve the problems of the organization.

6. To what degree do personnel have shared organizational values, a vision, focus, and goals? I think this is one of the most overlooked issues in

any organization, whether it's a fire department or a Fortune 500 company. The culture and expectations of the organization will often dictate its success. Shared values often start with the leader of the organization outlining his or her values to the group. Be willing to write them down and incorporate them in your organization's strategic planning process, and budget documents, posting values, value statements, and codes of conduct on the wall. That begins to create a shared expectation of how people within the organization should act, and it helps to develop a clear understanding of what you, the chief fire executive, expect. Create shared values for the organization through many different processes—labor–management committees, volunteer or association–management processes, strategic planning, project initiatives, team building, and community outreach programs. Values that are accepted and shared by the majority of the workforce help to drive the organization. These shared values will provide a mechanism for developing a shared vision, goals, and objectives.

A New Beginning

As you begin to build your team through your agenda for change and creation of a new culture, your organization will evolve to a new level of understanding, attitude, and value and know that you're embarked on a new beginning. This process is found to create mixed emotions. Old anxieties are often reactivated from the other two places we've already considered—transition and the neutral zone. People look at the new way of doing things as a gamble. Leading change is partly about providing a safety net so that your people feel comfortable enough to take a risk, knowing that if they fail, they'll survive.

I've found that when you begin this process and the majority of your personnel are experiencing a new beginning, successful change often depends on what I call the "four P's": purpose, picture, planning, and part. It's vital that you explain the purpose behind your change objective. Framing the purpose lets the action become real to the people involved; in many cases, there is no purpose until it has been clearly defined. You can strengthen that sense of purpose by painting a picture of how the outcome will look and feel. One way is to lay out a step-by-step plan, including what the ultimate outcome will look like. Remember, in the fire service, we're trained for fire ground operations, so we're a linear group. Teams are an essential part of the change process. They are also a factor that we often overlook. Providing this framing experience is essential for evolving your team's culture and embarking on the change process. Here is a useful Team Performance Model.

Team Performance Model

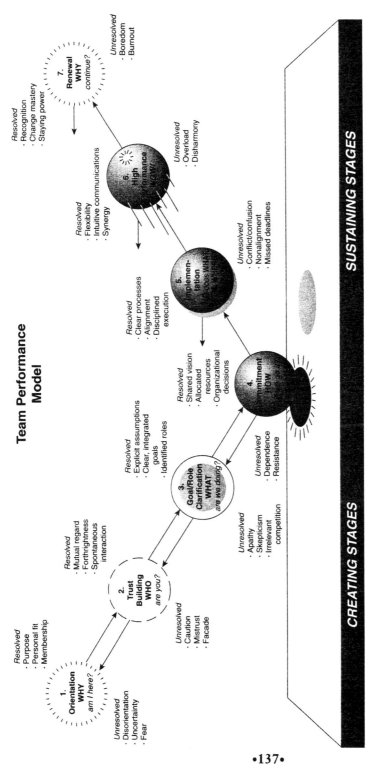

An excellent snapshot can be gained through such simple techniques as surveying your senior staff departmental leadership.

IDENTIFYING ORGANIZATIONAL PAINS

Organizational Growing Pains and Aging Pains are not only problems in and of themselves, but are symptoms of deeper systemic problems—they indicate that an organization's existing infrastructure no longer "works" and needs reorganization.

Ten Growing Pains

A = To a Very Great Extent   **B** = To a Great Extent
C = To Some Extent   **D** = To a Slight Extent
E = To a Very Slight Extent

1. A B C D E   People feel that "there are not enough hours in the day."
2. A B C D E   People spend too much time "putting out fires."
3. A B C D E   Many people are not aware of what others are doing.
4. A B C D E   People lack an understanding about where the organization is heading.
5. A B C D E   There are too few good managers.
6. A B C D E   People feel that "I have to do it myself if I want to get it done correctly."
7. A B C D E   Most people feel that meetings are a waste of time.
8. A B C D E   Plans that are made get little follow-up, so things just don't get done.
9. A B C D E   Some people feel insecure about their place in the organization.
10. A B C D E   The organization continues to grow, but staff levels have not kept up.

Scoring

11. _ _ _ _ _   Add the total number of responses in each column.
12. 5 4 3 2 1   Multiply the number on line 11 by the number on line 12, and record the result on line 13.
13. _ _ _ _ _   Result of line 11 times line 12.
14. _ _ _ _ _   Add the numbers on line 13, columns A–E, and place the result on this line.

Ten Aging Pains

A = To a Very Great Extent   **B** = To a Great Extent
C = To Some Extent   **D** = To a Slight Extent
E = To a Very Slight Extent

1. A B C D E   We know what's best for our customers; we don't need to ask them.
2. A B C D E   People are spending too much time covering their vested interests.
3. A B C D E   We're mired in bureaucracy and red tape.
4. A B C D E   On the surface, we're one big family, but there's a great deal of conflict below the surface.
5. A B C D E   People are increasingly unwilling to take risks.
6. A B C D E   Each manager has a kingdom; there's no sense of unified direction.
7. A B C D E   We have too many people doing the same thing and not enough people doing the right thing.
8. A B C D E   There are too many levels and too many people in the organization.
9. A B C D E   Performance isn't rewarded; it's who you know, not what you do.
10. A B C D E   Our managers don't know what they are doing and/or don't know what to do.

Scoring

11. __ __ __ __ __   Add the total number of responses in each column.
12. 5 4 3 2 1   Multiply the number on line 11 by the number on line 12, and record the result on line 13.
13. __ __ __ __ __   Result of line 11 times line 12.
14. _____   Add the numbers on line 13, columns A–E, and place the result on this line.

Interpretation

10–14	Everything okay
15–19	Some things to watch
20–29	Some areas that need attention
30–39	Some very significant problems
40–50	A potential crisis or turnaround situation

PROCESS SCAN

Patterns of Thinking Profile

Instructions: In each set, check the statement that more accurately reflects your own thinking. Be sure to respond to each of the sets.

1. ____ a. My mind scans over complex information and events, seeing new combinations.
____ b. My mind sorts out detail and makes good sense out of the confusion.

2. ____ a. I'm "high tech" and prefer precision and accuracy.
____ b. I'm "high touch" and value feelings and relationships.

3. ____ a. I sense the power of feeling tones, novel expressions, and swirling images.
____ b. I am better at handling numbers, factual information, and practical results.

4. ____ a. I paint the world with fresh ideas and explore new, uncharted horizons.
____ b. I pride myself on being organized and full of common sense.

5. ____ a. I thrive on changes, novelty, variety, and scrambled situations.
____ b. I prefer my world to be "cut-and-dried," stable, and predictable.

6. ____ a. I'm often criticized for being rigid, locked in, and matter-of-fact.
____ b. I get criticized for being too wishy-washy, theoretical, idealistic, and "far out."

7. ____ a. My mind roams freely over constantly changing landscapes.
____ b. My mind consistently sorts out and evaluates people, ideas, and projects.

8. ____ a. I generally see the "trees" instead of the "forest."
____ b. I am much more aware of the "forest" than the "trees."

Check only *one word* from each pair.

9. ____ a. Intuitive　　or　　Logical　　b. ____
10. ____ a. Organizer　　or　　Synthesizer b. ____
11. ____ a. Realistic　　or　　Impulsive　b. ____
12. ____ a. Detailer　　or　　Scanner　　b. ____　　SCORE: ____

13. ____ a. Spontaneous or   Methodical b. ____
14. ____ a. Visionary   or   Pragmatic b. ____
15. ____ a. Calculative or   Instinctive b. ____

Source: Kent Leyden & Associates

As we build any change initiatives, reorganization may be required. Reorganization may go against basic values that many in our departments hold dear. These proposed changes may contradict the way members believe things should be done, and as such, they often resist change for the following reasons:

1. They are afraid of the unknown.
2. They think things are fine and don't understand the need for change.
3. They are inherently cynical about change, particularly as the latest management technique.
4. They doubt there are effective means to accomplish major reorganizations.
5. They see conflicting goals in the organization (i.e., increasing resources to accomplish the change while cutting costs to remain viable).

Probably the best way to address these issues is through increasing communication. The chief should meet with all management staff to explain the reasons for the change, how in general it will be carried out, and where they can go for additional information. Detailed plans should be developed and communicated. Plans do change; understanding this, make sure you communicate changes and why they occurred. Encourage forums within the organization that allow employees to express their ideas, concerns, and frustrations. Whenever change occurs, something ends and something new begins. People experience a loss even though the change is positive. If the loss isn't acknowledged and managed, employees cannot be led into a new direction. The types of loss that employees experience include security, relationships, a sense of direction, ownership, and territory. People are more willing to make changes if they are led than if they are told what to do. Information will normally change their behavior, and they will probably respond to support and encouragement from their leadership. The more involved senior leaders are with their teams and each other, the easier the change will be. Those who create trusting relationships are more successful during periods of change. That's why the Team Performance Model reflects what I have seen my organizations go through during a change process, going repeatedly through the various levels of creating and sustaining performance as a team. As a leader, you must recognize that your leadership will help shape the direction the teams will take, and that the process will require understanding of and empathy with the work and emotions of the individual groups.

Last, but certainly not least, give each person a part to play in the plan. The more stakeholders you create as part of your change process, the more likely you'll be successful. People need a way to contribute or participate, and they want to have a role in relation to what others in the organization are doing.

Each person's role must be defined in regard to the outcomes, which you antic-ipate from a leadership perspective. Bringing every member of the organization into the process, defining their role, and providing clear direction and support will help to achieve a successful organizational change.

Building High-Performance Teams

As I pointed out previously, it's interesting that fire services can respond to an emergency, demonstrating extremely effective teamwork, yet form a very dys-functional team at the conference table. There the outcome is quite different when there is no purpose, motivation, understanding, and leadership. What can you, as fire chief, do to develop several high-performing teams creating opportunities, devising solutions for your most complex problems, and explor-ing future ideas and concepts? It often comes down to your ability to assess, and possibly change, the dynamics of the existing groups and then create an environment that allows them to develop into high-performance teams.

Students of business management are often introduced to Abraham Maslow's (1909–1970) ideas on motivation and people's needs. Dr. Maslow, a psychologist, created his now famous Hierarchy of Needs, which consisted of five needs:

Self-actualization
Esteem needs
Belonging needs
Safety needs
Psychological needs

Maslow's Hierarchy identifies a series of levels of satisfaction that either mo-tivate persons or discourage them from acting. While many people look at the Hierarchy as a stair-stepped approach that is continually escalating, it is actu-ally a dynamic process that is undergoing constant change. As individuals and as members of groups, all persons can function at different levels in the orga-nizational hierarchy, depending upon the actions of their superiors and/or other members of the group. A basic premise of Maslow's Hierarchy is that people cannot move to the next level until they have satisfied their needs at their current level. Similarly, groups will not have an adequate self-concept, a critical factor in performance, until they have worked through the lower lev-els and reached the esteem or group fulfillment level (the top level in Maslow's hierarchy).

No group can feel good about itself when it feels threatened. Each group must be given an opportunity to perform at its highest possible level. Never-theless, no group will function at its highest possible level if it is pushed back, held back, or otherwise prevented from ascending to the next level. Watch the groups that operate under your command the next time there is a personnel action, a lack of communication, or an unwanted or misunderstood policy. These groups will often move up and down the hierarchy based upon the forces and environmental impacts that are occurring.

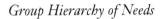

Group Hierarchy of Needs

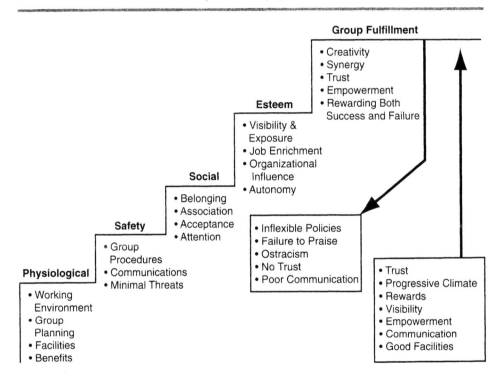

Chose a couple of work groups or fire companies to evaluate.

1. Where would you place each group on the hierarchy?
2. What factors have placed them at that satisfaction level?
3. What actions could you as a chief fire officer take to make their group more effective?

We often think of Maslow's Hierarchy of Needs as a theoretical aspect of employee psychology. But in fact, it has a practical application to group dynamics. You can see this in your own organization. Here are some examples. You go through contract arbitration; the organization wins and the firefighter's association loses. What happens to morale? Where do the employees immediately go on the Hierarchy of Needs? To the bottom, where they will focus on little else more than their own needs. You have to cut your budget by 15 percent, lay off ten firefighters, and close a fire station. Or your budget was increased, and you will be able to open a new fire station and purchase several pieces of equipment. Where do you think the majority of employees will be on the Hierarchy, and what challenges will you, as a leader, face as a result of each scenario? Events can drive the entire organization, or different groups within it, different levels within the Hierarchy of Needs. Understanding that is very important from a leadership perspective, as it will help you to determine what your challenges are and where you need to focus your time and energy.

The Framework of the High-Performance Programming Model

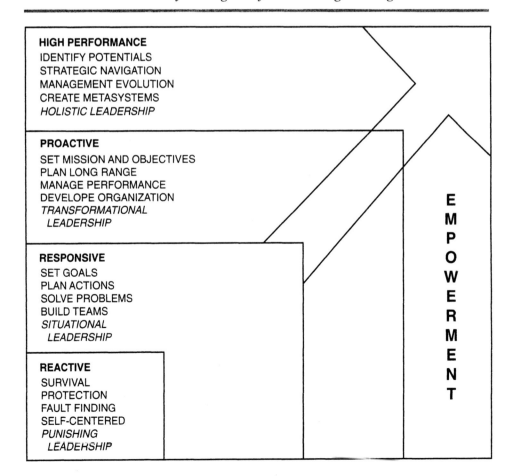

The second assessment tool is the framework for high-performance programming. Take the same groups that you used before and place them in the framework model.

1. Where would you place each group in the model?
2. What factors have placed them there?
3. What actions as chief can you take to move them to a higher level?

If the teams you selected were placed in the lower portions of either the group hierarchy or the high-performance model, they probably were exhibiting some dysfunctional behavior. What action can you take to help develop positive team dynamics and move them to a high level of understanding, acceptance, and performance?

Characteristics of a High-Performance Team

High-performance teams are seldom an accident. Rather, they are an articulation of leadership that stems from adherence to basic principles such as trust, meaningful responsibility, commitment, high standards, and the willingness to confront inadequate performance and reward exceptional performance.

In a group, several elements, both internal and external, contribute to the development of high-performance teams. These factors increase or decrease the ability of the team to perform as a group.

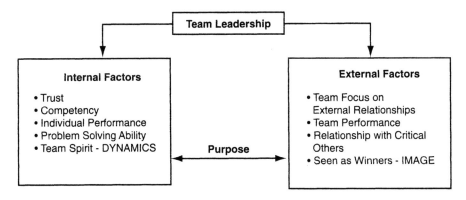

Focus on the Individual

Leading high-performance groups necessitates an understanding of the needs of individuals in a group context. Everyone in a group situation has three primary needs:

1. *Inclusion:* Every person entering a new group or belonging to an existing one needs to develop a viable role or identity within the group.

2. *Influence:* Every person in the group needs a certain amount of influence and control.
3. *Acceptance:* Every person in the group needs to feel a degree of personal acceptance.

These three factors reflect the basic human need for security and must be achieved in the development of any team. If you are a new member of a group, including the role of fire chief, you will struggle to satisfy those needs. When new groups are formed, they cannot fulfill their mission until the members learn to meet their own needs to some degree. The feeling of comfort in a new group situation can be enhanced by focusing on the purpose of the group and/or by providing activities designed to help the members gain an understanding of each other.

As individuals join groups, they bring with them their own frame of reference based on their own styles and experiences. When opposites work together, there can often be frustration if the individuals involved do not understand the differences in the way the other person approaches problems. Understanding that people approach situations from a different frame of reference often reduces disagreement and can actually become a learning experience.

Two of the most beneficial tools for understanding how individuals interact as part of a group are the Keirsey Temperament Sorter and Building High Performance Teams, by Richard Ross and Tom Isgar.

Individual Styles

The Keirsey Temperament Sorter accounts for the variations in cognitive styles that arise because people have different assumptions about the nature of the truth, human nature, and relationships. One of the most widely researched and used theories on psychological type/cognitive styles is the Myers-Briggs Type Indicator. This theory, based on the work of the Swiss psychologist Carl Jung, provides an understanding of individual differences based on preferences on four scales. The four Myers-Briggs types are:

1. *Introversion/Extroversion:* This scale indicates where people get their energy and focus their attention. Introverts focus on the inner world of ideas and concepts. Extroverts focus on the outer world of people and things.
2. *Sensing/Intuition:* This scale describes opposite ways of obtaining information from the environment. Sensing types trust and rely most on their five senses. Intuitive types trust and use their intuition, extrapolating meaning and relationship from their previous experiences.

3. *Thinking/Feeling:* This scale refers to how people reach conclusions, make decisions, and form opinions. Thinking types use logical, objective analysis. Feeling types weigh alternatives based on values: how important something is and how much people care about it.
4. *Judging/Perceptive:* This scale refers to the type of lifestyle individuals prefer and refers back to the previous two scales. Judging types tend to live more in the decision making mode and prefer planned, orderly lives. Perceptive types spend more time taking in information, and tend to like flexibility and spontaneity.

The Keirsey Temperament Sorter provides a means to identify sixteen characteristic styles. Each style includes preferences typically exhibited by people who fall into the different style preferences based upon the scores from the Temperament Sorter. The Keirsey Temperament Sorter, based upon the Meyers-Briggs Type Indicator Test (published by Consulting Psychologist, Palo Alto, CA), are distributed and researched by the Center for Applications of Psychological Type (P.O. Box 13807, Gainsville, FL 32604). The book *Please Understand Me*, by David Keirsey and Marilyn Bates, is distributed by the Prometheus Nemisis Book Company (P.O. Box 2748, Del Mar, CA, 92014). **The Leadership Dimension of Communication and Problem Solving.** In the Keirsey Temperament Sorter, the E (extrovert) and I (introvert) preferences indicate where persons get their energy and zest for life. E's are stimulated by the outside environment—by people, places, and things. I's are stimulated from within—by their own world of thoughts and reflections. E's are energized by what is going on around them (*parties prove this*) and do their best work externally by taking action. They prefer to communicate openly and freely. I's are energized by their personal experiences and inner resources, and they do their best work internally by reflecting and thinking before taking action. They prefer not to communicate openly and freely until they know and trust someone (*have you ever tried to stimulate a group interaction with a group of I's who do not know one another?*). E's focus on broad implications, and I's focus on single subjects in depth. Both are required for effective problem solving and can be effectively used if their partners understand their differences.

Partners with the same preference may lack either a broad or an in-depth perspective in their analysis, which presents a different challenge. How do they achieve balance in their problem solving?

E's tend to think and problem-solve out loud; you know that they are problem-solving because you can see their lips moving and almost hear them thinking. E's often say, "Just let me talk long enough and I will figure out the answer." I's typically problem-solve differently. They require a short period of quiet time to think about and reflect on the problem before they can discuss it. I's typically say, "If I can just get some quiet time to hear myself think, I can figure out the answer."

A significant challenge occurs when you mix E's and I's together in problem solving. The E's immediately start talking out loud and the I's cannot hear themselves think. As a result, the I's may not interact in the problem-solving process at all.

Two E's (*Is anyone listening?*) or two I's (*Is anyone talking?*) also have challenges when communicating and problem solving.

Extrovert (E)	Introvert (I)
An E's essential stimulation is from the environment—people and things	An I's essential stimulation is from within—thoughts and reflections
Energized by other people, external experiences	Energized by inner resources, internal experiences
Does best work externally in action Interests have breadth	Does best work internally in reflection Interests have depth
Usually communicates freely—expressive	Usually reserved in communication until he or she knows and trusts a person
Acts, perhaps reflects, acts	Reflects, perhaps acts, reflects
Thinks best when talking with people	Thinks best when alone; shares with others when clear about what he or she believes
Usually takes the initiative in making contact with other people	Usually lets other people initiate contact
Has broad friendships with many people—gregarious	Has a few deep friendships—intimacy
Prefers to talk and listen	Prefers to read and write

The Leadership Dimension of Planning

In the Keirsey Temperament Sorter, the Sensing and iNtuiting functions are ways that people prefer to perceive and take in information. The Sensing function takes in information by way of the five senses and likes to look at specific pieces of that information, deal with known facts, and live in the present, enjoying what exists. The iNtuiting function also takes in information via the five senses but then adds a sixth sense—*a hunch or intuition*. Most N's will state, "I make my worst decisions when I go against my intuition." N's are very conceptual and prefer to look at overall patterns and relationships. They like to deal with broad concepts or possibilities, and they plan for the future. They enjoy anticipating what might be.

As a result of broad preferences, S's and N's tend to approach planning differently. N's prefer the broad, overall conceptual view, like to work with future possibilities, and are comfortable with envisioning processes. N's like to define *where* the organization is going and the possible attributes, conditions, and outcomes that it may seek to obtain. S's prefer step-by-step pragmatic planning based upon what can be accomplished today. They are most comfortable developing strategies, steps, and action plans to achieve certain goals. They prefer to define how the organization is going to achieve its goals. If N's conceived of putting a man on the moon, S's devised the systems and hardware to make it happen. Obviously, a good plan requires both perspectives—long-range conceptual goals (*where*) and pragmatic strategies and action plans (*how*). It is important to understand which strengths and preferences each partner brings to the planning process.

Partners with the same preferences may not focus on one of the two critical planning components—either the where or the how.

Sensing (S)	Intuiting (N)
The S function takes in information via the five senses	The N function processes information via both the five senses and a *sixth sense—hunch*
Looks at specific parts and pieces	Looks at patterns and relationships
Deals with facts	Deals with possibilities
Lives in the present, enjoying what is there	Lives for the future, anticipating what might be
Trusts experience	Trusts theory more than experience
Tends to be seen as realistic	Tends to be seen as imaginative
Likes to apply reliable, proven solutions to problems	Likes problems that require new solutions
Likes the concrete	Likes the abstract
Learns sequentially, step by step	Learns by seeing connections—jumps in anywhere, leaps over steps
Tends to be good at precise work	Tends to be good at creating designs

Characteristics Frequently Associated with Each Type

Sensing Type		Intuitive Type	
ISTJ	**ISFJ**	**INFJ**	**INTJ**
Serious, quiet, earn success by concentration and thoroughness. Practical, orderly, matter-of-fact, logical, realistic and dependable. See to it that everything is well organized. Take responsibility. Make up their own minds as to what should be accomplished and work toward it steadily, regardless of protests or distractions.	Quiet, friendly, responsible and conscientious. Work devotedly to meet their obligations. Lend stability to any project or group. Thorough, painstaking, accurate. Their interests are usually not technical. Can be patient with necessary details. Loyal, considerate, perceptive, concerned with how other people feel.	Succeed by perseverance, originality and desire to do whatever is needed or wanted. Put their best efforts into their work. Quietly forceful, conscientious, concerned for others. Respected for their firm principles. Likely to be honored and followed for their clear convictions as to how best to serve the common good.	Usually have original minds and great drive for their own ideas and purposes. In fields that appeal to them, they have a fine power to organize a job and and carry it through with or w/o help. Skeptical, critical, independent, determined, sometimes stubborn. Must learn to yield to less important points in order to win the most important.
ISTP	**ISFP**	**INFP**	**INTP**
Cool with onlookers—quiet, reserved, observing and analyzing life with detached curiosity and unexpected flashes of original humor. Usually interested in cause and effect, how and why mechanical things work, and in organizing facts using logical principles.	Retiring, quietly friendly, sensitive, kind, modest about their abilities. Shun disagreements. Do not force their opinions or values on others. Usually do not care to lead but are often loyal followers. Often relaxed about getting things done, because they enjoy the present moment and do not want to spoil it by undue haste or exertion.	Full of enthusiasms and loyalties but seldom talk of these until they know you well. Care about learning, ideas, language and independent projects of their own. Tend to undertake too much, then somehow get it done. Friendly, but often too absorbed in what they are doing to be sociable. Little concern with possessions or physical surroundings.	Quiet and reserved. Especially enjoy theoretical or scientific pursuits. Like solving problems with logic and analysis. Usually interested mainly in ideas, with little liking for parties or small talk. Tend to have sharply defined interests. Need careers where some strong interest can be used and useful.
ESTP	**ESFP**	**ENFP**	**ENTP**
Good at on-the-spot problem solving. Do not worry, enjoy whatever comes along. Tend to like mechanical things and sports, with friends on the side. Adaptable, tolerant, generally conservative in values. Are best with real things that can be worked, handled, taken apart, or put together.	Outgoing, easygoing, accepting, friendly, enjoy everything and make things more fun for others by their enjoyment. Like sports and making things happen. Know what's going on and join in eagerly. Find remembering facts easier than mastering theories. Are best in situations that need sound common sense and practical ability with people as well as things.	Warmly enthusiastic, high-spirited, ingenious, imaginative. Able to do anything that interests them. Quick with a solution for any difficulty and ready to help anyone with a problem. Often rely on their ability to improvise instead of preparing in advance. Can usually find compelling reasons for whatever they want.	Quick, ingenious, good at many things. Stimulating company, alert and outspoken. May argue for fun on either side of question. Resourceful in solving new and challenging problems, but may neglect routine assignments. Apt to turn to one new interest after another. Skillful in finding logical reasons for what they want.
ESTJ	**ESFJ**	**ENFJ**	**ENTJ**
Practical, realistic, matter of fact, with a natural head for business or mechanics. Not interested in subjects they see no use for, but can apply themselves when necessary. Like to organize and run activities. May make good administrators, especially if they remember to consider others' feelings and points of view.	Warm-hearted, talkative, popular, conscientious, born co-operators, active committee members. Need harmony and may be good at creating it. Always doing something nice for someone. Work best with encouragement and praise. Main interest is in things that directly and visibly affect people's lives.	Responsive and responsible. Generally feel real concern for what others think or want, and try to handle things with due regard for the other person's feelings. Can present a proposal or lead a group discussion with ease and tact. Sociable, popular, sympathetic. Responsive to praise and criticism.	Hearty, frank, decisive, leaders in activities. Usually good in anything that requires reasoning and intelligent talk, such as public speaking. Are usually well informed and enjoy adding to their fund of knowledge. May sometimes appear more positive and confident than their experience in an area warrants.

Utilizing the Keirsey Temperament Sorter from a team-building perspective will provide a personality portrait of team members and is an enjoyable exercise to promote positive group dynamics. In every organization that I have taken over, I have used the Kiersey Temperament Sorter to identify the traits of my senior staff. It's a very easy instrument to use and takes less than thirty minutes. It has an uncanny ability to identify the individual traits that your leadership team will bring to the table. It's an excellent way to understand the psychological and cognitive styles of your senior leadership team in a very short time, and it will give you some perspective on how they process information, which will be very beneficial.

The Importance of Task and Process Characteristics

In order for a group of individuals to become a team, they must achieve a unifying goal or accomplish a task. The term "group process" refers to how well members interact with each other to accomplish the task. Individuals differ in their needs to focus on the task or process and in their skills in each of these areas. Attention to both is critical to team success. Following are task and process roles that must be played for effective team interaction.

Task Roles

1. *Problem Definition:* defines the group's problem.
2. *Seeks Information:* requests factual information about group problems or procedures, or asks for clarification of suggestions.
3. *Gives Information:* offers facts or general information about group problems or procedures, or clarifies suggestions.
4. *Seeks Opinions:* asks for the opinions of others relevant to the discussion.
5. *Gives Opinions:* states beliefs or opinions relevant to the discussion.
6. *Tests Feasibility:* questions reality, checks practicality of suggested solutions.

Process Roles

1. *Coordination:* clarifies a recent statement and may relate it to another statement in such a way as to bring both together; reviews proposed alternatives.
2. *Mediating/Harmonizing:* intercedes in disputes/disagreements and attempts to reconcile them; highlights similarities in views.

3. ***Orienting/Facilitating:*** keeps the group on track, points out deviations from agreed-upon procedures, direction of group discussion; helps the group process by suggesting other procedures to make the group more effective.

4. ***Supporting/Encouraging:*** expresses agreement with others' ideas, verbally supports others.

One of the best tools I have seen to determine leadership characteristics is an instrument developed by Richard Ross and Tom Isgar entitled Building High Performance Teams. It was introduced to me as part of my graduate course work at Regis University, Denver, in a class on change and leadership. I have found it to be useful in determining who in the group is task oriented versus process oriented. Through measurement of four categories—innovator, achiever, organizer, and facilitator—a team leader can identify the strengths of each team member in order to get the most from the team.

Innovators Like New Ideas

- Brainstorming is great
- Crazy ideas—the more, the better
- Stimulate thinking by challenging the status quo
- A better way is the only way
- A risk-free Atmosphere is a must
- Time is not a big factor

Achievers Like to be Directed

- Goal-directed—when, where, and how
- Can overcommit to too many projects
- Need to get things done to feel good
- Momentum is created by many projects
- Action-oriented: get things done, move on to the next project
- Reports on the status to keep track of all the projects
- Let's celebrate when we finish a project

Organizers Like to Follow the Steps

- Resources are well defined
- Preparation is a must
- All steps must be followed
- Time efficiency is critical to the organizer's organizational plan

Facilitators Like Good Process and Consensus Building

- Clear expectation of the role/purpose of each member
- Designs —how to make it happen

- Sensitive to feelings of the team
- Therefore, involves more of the strategies of all members to build consensus

Measuring Your Leadership Style

Ross and Isgar's Building High Performance Teams survey evaluates sixty items that describe characteristics of leaders in organizations. A variety of situations common to leadership have been included to cover a wide range of characteristics and, thereby, to provide you with meaningful information about yourself as a leader. You can also use this survey to determine the leadership characteristics of your senior staff and station officers. It is an interesting exercise, can be insightful, and can help to open a meaningful dialogue.

Each major category (e.g., "Innovation") is represented in fifteen different situations. The four alternatives to each situation differ. Therefore, read all four alternatives before answering so that you can select the alternatives *most* and *least* characteristic of you. There are no right or wrong answers. The best answer is the one that is most descriptive of you. Therefore, answer honestly, since only realistic answers will provide you with any useful information about yourself.

Instructions: From each set of four alternatives, select the one that is most characteristic of you. Place the letter for that item on the scale at the point which reflects the degree to which that item is characteristic of you. Then, select the alternative that is least characteristic of you and place its letter at the appropriate point on the scale. Once you have found the most and least characteristic alternatives, enter the letters of the remaining alternatives within this range according to how characteristic each alternative is of you. Do not place alternatives at the same point on the scale (no ties).

For example, on a given set of four items you might answer as follows:

1. a) Imaginative   c) Slightly Competitive
   b) Sensitive   d) Rational

Completely Characteristic   **Completely Uncharacteristic**

```
:__:__:D:C:A:__:__:B:__:__:__:__:
 10  9  8  7  6  5  4  3  2  1  0
```

Place the letter of the "characteristic" on the scale above the number to reflect your order of importance.

Building High Performance Teams: Leadership Roles

1. a) Imaginative
    b) Sensitive

    c) Slightly Competitive
    d) Rational

Completely Characteristic   **Completely Uncharacteristic**

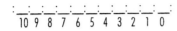

2. a) Prudent
    b) Listener

    c) Pragmatic
    d) Risk-taker

Completely Characteristic   **Completely Uncharacteristic**

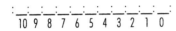

3. a) Results-oriented
    b) Concerned for feelings of others

    c) Creative
    d) Analytical

Completely Characteristic   **Completely Uncharacteristic**

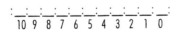

4. a) Situationally-sensitive
    b) Hard Worker

    c) Organized
    d) Innovative

Completely Characteristic   **Completely Uncharacteristic**

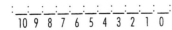

5. **Emphasizing**

    a) Getting Results
    b) Human Interaction

    c) Logical Thinking
    d) Ideas and Innovation

Completely Characteristic   **Completely Uncharacteristic**

6. **Producing Results by**

    a) Getting People to Work Together
    b) Doing My Part

    c) Motivating Other People
    d) Establishing Helpful Systems & Procedures

Completely Characteristic   **Completely Uncharacteristic**

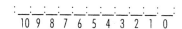

7. **Tending to Focus on**

a) What Is New That Could be Done   c) How the Work Is Being Done
b) The Caliber of What Is Being   d) What Is Being Done

Completely Characteristic   **Completely Uncharacteristic**

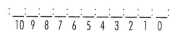

8. **Being Described as**

a) Highly Dedicated to My Work   c) Thoughtful and Precise
b) People-oriented   d) Charismatic and Enthusiastic

Completely Characteristic   **Completely Uncharacteristic**

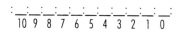

9. **In Meetings Tending to**

a) Talk   c) Listen for Underlying Motives &
b) Listen Impatiently   Potential Conflict
                                      d) Listen Critically for What May Not
                                         Work

Completely Characteristic   **Completely Uncharacteristic**

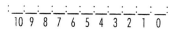

10. **In a Work Group**

a) Wanting to Keep People Working Together Harmoniously
b) Looking at What We Could Be Doing Differently
c) Wanting to Do the Job Efficiently
d) Like to Keep People Focused on the Task

Completely Characteristic   **Completely Uncharacteristic**

10 9 8 7 6 5 4 3 2 1 0

11. Valuing Information That Is

a) About Technical Concerns
b) About Goals and Policies
c) About Opportunities
d) About Others' Beliefs on Issues

Completely Characteristic   **Completely Uncharacteristic**

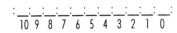

10 9 8 7 6 5 4 3 2 1 0

12. Using Free Time to

a) Think About New Methods, Procedures, Controls, Etc.
b) Seek Out Information from Others, Sift Rumors, Interact
c) Create a New Project or Revamp/Change Ongoing Work
d) Complete More Work

Completely Characteristic   **Completely Uncharacteristic**

10 9 8 7 6 5 4 3 2 1 0

13. In Making a Decision

a) Look for the Right Answer
b) Use Group Input
c) Act Quickly
d) Consider Precedents

Completely Characteristic   **Completely Uncharacteristic**

10 9 8 7 6 5 4 3 2 1 0

14. In a Conflict Situation

a) Use Conflict to Pressure Subordinates
b) Become Uncomfortable
c) Ignore it, or Fight Based on Rule and Policies

Completely Characteristic   **Completely Uncharacteristic**

10 9 8 7 6 5 4 3 2 1 0

15. Excelling By

a) Getting Things Done
b) Accomplishing Things with Systems and Procedures
c) Creating New Venture
d) Working with Groups to Accomplish Tasks or Get Agreement

Completely Characteristic   **Completely Uncharacteristic**

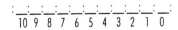

10 9 8 7 6 5 4 3 2 1 0

LEADERSHIP STYLES

Scoring Key

To score the inventory, merely transfer the rankings you assigned to the appropriate item number below, and then add the columns. If possible, we suggest you pair up with another person to read off the rankings.

	Innovator	Achiever	Organizer	Facilitator
1.	a. _____	c. _____	d. _____	b. _____
2.	a. _____	c. _____	d. _____	b. _____
3.	a. _____	c. _____	d. _____	b. _____
4.	a. _____	c. _____	d. _____	b. _____
5.	a. _____	c. _____	d. _____	b. _____
6.	a. _____	c. _____	d. _____	b. _____
7.	a. _____	c. _____	d. _____	b. _____
8.	a. _____	c. _____	d. _____	b. _____
9.	a. _____	c. _____	d. _____	b. _____
10.	a. _____	c. _____	d. _____	b. _____
11.	a. _____	c. _____	d. _____	b. _____
12.	a. _____	c. _____	d. _____	b. _____
13.	a. _____	c. _____	d. _____	b. _____
14.	a. _____	c. _____	d. _____	b. _____
15.	a. _____	c. _____	d. _____	b. _____
	TOTAL	TOTAL	TOTAL	TOTAL

The High Performance Team

Teams who reach a high performance level do so through effective leadership and realizing the potential of their collective experience and expertise. Whatever methodology you use to get there, high performance teams exhibit common characteristics. The following describes these characteristics.

Characteristics of High Performance Teams

Individual Level

- High personal commitment to the group
- High level of trust among members
- Involvement with the group inspires each individual's personal best
- High level of personal development
- Fun and excitement

Group Level

- Internalized purpose and mission is a basis of action
- Results driven . . . keen focus on making a difference
- Effective utilization of members' skills and abilities (diversity of skills, points of view, and values)
- Open communication . . . effective norms for surfacing and working through differences
- High standards of excellence
- Smooth task and process flow within the group
- Open expressions of appreciation, recognition and caring
- Optimal communication and exchange with the "outside world"
- Willingness to experiment and try new ways of doing things . . . flexibility and versatility

Source: Adams (1984).

As you've realized, potential often begins with gaining an understanding of the people in the group. Although we have discussed two written instruments, there are myriad others that can be used to gain a perspective on team and personality traits including rope courses, initiative games, and adventure activities. Whatever you decide to use, team building starts with understanding how others process information and what motivates them to participate.

Leaders realize that once visions exist, they must act to implement them. The leader's actions permeate all levels of the organization—from the newest recruit to the most senior staff. Leaders also recognize that when people step outside of their comfort zone in order to create a new organizational vision, success depends on the leader's ability to institutionalize the change—within the culture, the values, the people, and the organization as a whole. Organizations, like individuals, need motivation to succeed. Understanding individual leadership and personality styles will be an important determinant of your ability to form your teams and motivate them to do great things.

Compontents of a High-Performance Team

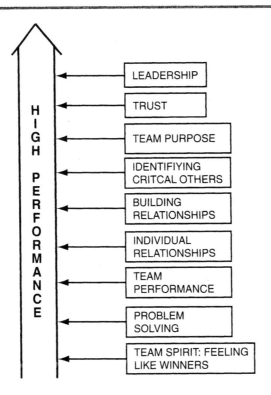

HIGH PERFORMANCE

- LEADERSHIP
- TRUST
- TEAM PURPOSE
- IDENTIFIYING CRITCAL OTHERS
- BUILDING RELATIONSHIPS
- INDIVIDUAL RELATIONSHIPS
- TEAM PERFORMANCE
- PROBLEM SOLVING
- TEAM SPIRIT: FEELING LIKE WINNERS

What Lies Ahead

There's never been a time in the history of the fire service that is more exciting than the present. While it is difficult to predict what the future will be for the fire service, several critical factors are emerging that will revolutionize our industry. The increasing diversity of our employees, the way they think, their values, the services they provide, the expectations of our communities—all these will transform the fire service in a short period of time. The way we lead and manage people in the future will also change, and this may present our biggest challenge as our leadership styles and abilities change to match the future demands of our profession.

A Diversifying Workforce

Suppose that, in twenty years, 40 to 50 percent of our workforce consists of women and minorities. Think about how much change that will bring to our industry. The change will include not only the way we look, but also the way we approach our problems and our industry.

In addition, for the first time in the history of the fire service, and in the history of our country, four generations will be in the workforce at the same time. Each generation has a unique set of values and expectations. This can often cause conflict in organizations because the values and expectations of the different generations can collide. Couple that with the increasing diversity of the workforce versus the traditional makeup of the fire service, and you can see that we are about to enter a workforce revolution—involving diversity, value structures, and employee expectations. All will challenge the leadership to blend this new mosaic.

Must research has been conducted and many books have been written regarding the generational differences that exist today. The impacts of these differences on society reflect the shared characteristics and similar interests that can be found in each generational group. The four generations in our workforce are:

The veterans	1922–1943
The baby boomers	1943–1960
Gen-X'ers	1960–1980
Nexters	1980–2000
?	2000–

Why is this important? Each generation brings certain values, traits, and expectations to the workforce. Depending upon the makeup of your workforce, the influence of each generation on your organization will be in a continual state of transition. As workers retire, new personnel are hired, and the organization grows or shrinks; each of these events impacts the culture of the organization. The culture of the organization that you manage today will be very different in ten years. Why? Very simply, we are in the midst of a generational wave, in which one generation, the veterans, is leaving and a new one, yet to be named, is emerging. The result is a value shift in the organization. What about the next wave, when the baby boomers start to exit the workforce in large numbers?

As each of these generations moves through time, its members experience similar events that help to shape the way they think about and perceive things. Those formative events are often significant national, emotional, historical events that become a permanent and defining memory for the group. Such experience helps to shape the basic characteristic, values, work ethic, and outlook of the generations involved. Although not everyone will fit neatly into any generational grouping, the profiles of these groups are surprisingly accurate.

A Profile of Four Generations

Veterans 1922–1943	Baby Boomers 1943–1960	Gen X'ers 1960–1980	Nexters 1980–2000	2000–?
Also Known As				
Traditionalists	Boomers	X'ers	Millennials	
GI's mature	Postwar generation	Twenty-somethings	Generation Y	
World War II generation	Vietnam generation	Thirteeners	Generation 2001	
The silent generation	Sixties generation	Baby busters	Nintendo generation	
Seniors	Me generation	Post-boomers	Generation net	
The builder generation			Internet generation	
What Shaped Them				
Patriotism	Prosperity	Watergate, Nixon resigns	Computers	September 11
Families	Children in the spotlight	Latchkey kids	Oklahoma City bombing	War on Terrorism
The Great Depression	Television	Stagflation	*It Takes a Village*	War in Iraq
World War II	Suburbia	Single-parent home	TV talk shows	
New Deal	Assassinations	MTV	Multiculturalism	
Korean War	Vietnam	AIDS	Girls' movement	
Golden age of radio	Civil rights movement	Computers	McGuire and Sosa	
Silver screen	Cold war	*Challenger* disaster		
Rise of labor unions	Women's liberation	Fall of Berlin Wall		
School and church	The space race	Wall Street frenzy		
	Kent State killings	Persian Gulf war		
	Economic affluence	*Glasnost, perestroika*		
	Education and technology	*Roe* v. *Wade*		
Core Values:				
Dedication/sacrifice	Optimism	Diversity	Optimism	
Honor	Team orientation	Thinking globally	Civic duty	
Adherence to rules	Personal gratification	Balance	Confident	
Hard work	Health and wellness	Technoliteracy	Achievement	
Law and order	Personal growth	Fun	Sociability	
Duty before pleasure	Youth	Informality	Morality	
Respect for authority	Work	Self-reliance	Street smarts	
Conformity	Involvement	Pragmatism	Diversity	
Frugality				
Their View on the World:				
Outlook: Practical	Optimistic	Skeptical	Hopeful	
Work ethic: Dedicated	Driven	Balanced	Determined	
View of authority: Respectful	Love/hate	Unimpressed	Polite	
Leadership by: Hierarchy	Consensus	Competent	Pulling together	
Relationships: Personal sacrifice	Personal gratification	Reluctant to commit	Inclusive	
Turnoffs: Vulgarity	Political incorrectness	Cliché, hype	Promiscuous	

Source: Adapted from Zemke et al. (2000) and McIntosh (1995).

Defining Events

1930s	Great Depression Election of Franklin D. Roosevelt
1940s	Pearl Harbor D-Day Death of Franklin D. Roosevelt VE Day and VJ Day Hiroshima and Nagasaki atomic bombings
1950s	Korean War TV in every home McCarthy House Committee on Un-American Activities hearings Rock 'n roll Salk polio vaccine introduced
1960s	Vietnam War Election of John F. Kennedy Civil rights movement Assassinations of John F. Kennedy and Martin Luther King, Jr. Moo landing Woodstock
1970s	Oil embargo Resignation of Richard Nixon First personal computers Women's rights movement
1980s	*Challenger* explosion Fall of Berlin Wall Killing of John Lennon Election of Ronald Reagan
1990s	Persian Gulf War Oklahoma City bombing Death of Princess Diana Clinton scandals

Source: Adapted from Zemke et al. (2000).

As we move into the future and try to balance the needs of all of our employees, the role of leadership will become increasingly complex. At the same time that we are trying to find this balance and develop strategies to accommodate the different learning styles and lifestyle desires of the workforce of the future, our relationships with labor groups and volunteer associations are becoming strained. I believe that the next ten to fifteen years will bring a cyclical relationship between labor and management as we work to find a balance.

A cooperative labor–management partnership that deals with the changes in race, gender, ethnicity, worker generations, and shifts in our industry will be a must. This will help us prepare to meet the challenges involved in recruiting, retaining, and training quality people, diversifying our workforce, and leading the revolution in the fire service. It will force both labor and management leaders to change their approach to each other. The leaders of today and those of the future must strive for empowerment and inclusion. For both management and labor, this will mean becoming more global in their perspectives and collaborative in their approach.

The next several years will be very interesting from a labor–management standpoint. In the 1990s, the International Association of Fire Chiefs (IAFC) and the International Association of Firefighters (IAFF) worked on a collaborative effort called the Labor/Management Fire Service Leadership Partnership. Since 2000, they have cooperatively conducted Fire Service Leadership Partnership workshops throughout the country. These two-part workshops designed to promote cooperative labor–management relationships have paid dividends for many fire departments. Traditionally, they have been cosponsored by state fire chief and firefighter associations. The workshops focus on four goals: starting productive communications, understanding the cooperative process, understanding the benefits of the cooperative process, and providing tools to enhance partnerships. The participants engage in interactive discussions and exercises with their labor–management partners to gain a better understanding of each other and the pressure each faces from their respective superiors or constituents. These provide a way to understand the roles and relationships of labor and management and to create effective strategies at the local level. Hopefully, this will eliminate the conflicts of the past between fire chiefs and labor leaders, including votes of no confidence, picketing, and retribution from chief fire officers. The relationship between volunteer firefighters and local fire chiefs has also changed dramatically, as have the roles and responsibilities of the volunteer fire service. The demands placed on volunteers are becoming enormously time-consuming and stressful, leading to a whole new dynamic for the local chief fire officer who leads and manages a volunteer organization. As we move into the next decade, our increasingly diversified workforce will have a dramatic impact on our relationships with local labor leaders, local volunteer association presidents, and labor and management in general. At the same time, we have to understand how the changing workforce will affect the day-to-day dynamics of our departments.

Use of Technology

The technology available today—such as sprinkler systems and early warning devices—allows us essentially to eradicate fire in the United States. However, due to politics and differing agendas from the various groups in the fire service and lack of support from our counterparts in the construction industry, it will take leadership for this goal to be achieved in the foreseeable future. During the next ten to fifteen years, as the technology becomes more widely accepted and its costs are reduced, more fixed fire suppression retrofits will be incorporated in existing residential and commercial structures. The equipment and technology available for our field personnel will also improve, allowing their fire suppression activities to become more effective and increasing their safety dramatically.

The fire service of the next decade, due to the technology applications that will become available, will look very different than it does today. We will continue to see a decline in fire suppression activities and an increase in activities related to EMS, specialized rescue, emergency management, homeland security, and community outreach. In fact, by the next decade, I predict that fire suppression–related calls in many jurisdictions will drop below 10 percent. As the technology of building design and fire equipment improves, we will see a fire service in the very near future that looks, dresses, deploys, and performs very differently than it does today. One of the challenges that we as leaders will face is to help our people understand that these changes will occur in the near future and how they will affect their roles and responsibilities.

In fact, if we project the trend line of the American fire service over the past twenty years, and factor in technology, fixed fire protection, and better education, what we may see is a significant reduction in fire deaths and fire losses in the next twenty years. Therefore, our service may become a department of emergency services and not a fire department in the future. Fire will be just one aspect of a multitasked agency. Such a reinvention of our industry will require a mindset change in the culture of the fire service. The leaders of today must begin to plant the seeds for that change in the leaders of the future.

Changing Roles of the Future

Since the tragedy of September 11, 2001, I believe people view the fire service in a much different way. This service has been, and will become, an even greater partner in the planning and response to terrorist-related events. To do this, the fire service must partner at the local level much more effectively than ever before. We have to establish ourselves as credible, knowledgeable, involved practitioners and participate in things that go far beyond our normal roles as fire/EMS service providers. As a result of September 11, the public expects an integrated and cooperative response involving all government agencies. That changes what the fire service will do and how it will do it in the future. As chief fire officers, it is our duty to begin to prepare our organizations for these changes.

Funding

Over the past decade, we have seen many fire service organizations struggle to obtain sufficient funding for the variety of services they now offer. Today the fire service needs to rethink its approach to funding. We must provide performance measures that demonstrate how we add value to our communities. We must show how we are vital to the quality of life in the jurisdictions we serve, and why it is important for the safety and security of the residents we protect to have adequate fire and EMS services. To achieve this, many fire service leaders today must change their perspectives on what our industry is and what it is to become. Today we are still focused on basic performance measures such as response times, number of personnel, and fire loss. What we need is much more in-depth, quantifiable information on our relationship to the community and the value-added of the services we provide. The true cost of fire protection includes not only property saved, but also the cost of insurance and the value of saving a life, as well as explicit performance measures and outcome-based performance matrices that quantify the quality of services provided in relation to the funding that is available. Leadership in the future will need to look not only at how the fire service is funded, but also at the way we demonstrate how that funding adds value to the community and achieves the community-expected level of performance.

Our Time to Lead

Our profession, with its historical roots dating back over a 1000 years to the knights of Malta and the Knights of St. John, is one of tradition, dedication, and commitment. The North American fire service, originated in local communities and in the tradition of neighbor protecting neighbor. From these humble beginnings, the fire service has grown to what it is today and what it will become in the future—a group of highly trained volunteer and career men and women who respond to a variety of hazards with a common mission: to protect and to provide service over self. With the latest technology in firefighting and in emergency medical response to Hazmat mitigation, the fire service is charged with protecting our homes, our citizens, our nation, and our economy from a variety of threats. Those who wear the badge are motivated by the same sense of duty and pride that has inspired generations of firefighters. Today, the fire service continues to stand ready to work for the safety and security of all.

While the structure and organization of our local emergency response systems are always changing, one constant remains: the role of the fire service and its response when someone calls for help. In almost every emergency situation, whether a car wreck, a Hazmat spill, a medical emergency, a structural collapse, a residential fire, a natural disaster, or a terrorist event, it's the local fire service that is first on the scene and often the last to leave. The fire service has the ability, the equipment, and the personnel to offer the greatest degree of protection to community residents in times of national peril. On September 11, 2001, the world watched the fire departments in New York City, Arlington, Virginia, and the fields of Pennsylvania respond as we always do, and through their selfless actions, many firefighters that day lost their lives. We also know that through their actions, their commitment, and their courage, many lives were saved. This is the mission of the fire service: to be on the front line of defense, whether it's for our nation, our community, or the neighbor down the street. The fire service of the future will build upon the progress of the past several decades. That future will be tied to improvements in technology such as advanced imaging cameras, new radio communication systems, and improved personnel protective technology that will make firefighting practices safer and more effective. In the fire service of the future, increasing and improving fire prevention programs will be vital. Early detection and fixed fire extinguishing systems will be widely utilized, in residential, commercial, and industrial buildings, to reduce or eradicate fires. Widespread implementation of such technology will significantly reduce life and property loss, and will change the face of the fire service and the way it is deployed. As we incorporate this technology, our field commanders will be able to deploy their resources more effectively and allow fire department leaders to analyze service delivery needs much more effectively. For many fire services, it will also improve their ability to measure customer service relationships, improve productivity, and reduce expenses. Technology will help us to capture data so that we can reduce the dangers to our firefighters, ultimately saving the lives and property of the citizens we are charged to protect.

In the twenty-first century, we must be ready to build upon the important accomplishments of the past thirty years. In that time the fire service has grown exponentially; it is no longer considered just a local resource. Today, the fire service is considered an integral part of the effort to achieve national security, reduce regional economic liability, and maintain the local quality of life. If new threats emerge and new technologies are applied as our workforce diversifies, the leaders of the next generation will know that the challenge of the future must be met with innovation tempered by the realities of the past three decades. The leaders of tomorrow will meet this challenge by establishing partnerships with local, state, and national officials, carefully coordinating their efforts to improve the safety and security of our firefighters and our citizens. That is what we do, and that is why it is so important to understand that the badge we wear is truly a commitment and a promise to deliver to those we serve.

A symbol is a promise.

Source: Courtesy of Linda Stone.

SELECTED READINGS

Abrashoff, Michael D. *It's your ship*. New York: Warner Books, 2002.

Bay, Tom. *Look within or do without*. Franklin Lanes, NJ: Book Mart Press, 2000.

Bennis, Warren. *Why leaders can't lead: The unconscious conspiracy continues*. Los Angeles: University of Southern California Graduate School of Business, January 1990.

Bennis, Warren. "21st Century Leadership." *Executive Intelligence* May 1991, pp. 2–4.

Bennis, Warren. *On becoming a leader*. New York: Perseus Books, 2003.

Block, Peter. *The empowered manager*. San Francisco: Jossey-Bass Inc., Publishers, 1987.

Block, Peter. *Stewardship*. San Francisco: Berrett-Koehler Publishers, Inc., 1996.

Bolman, Lee G. and Deal, Terrence E. *Modern approaches to understanding and managing organizations*. San Francisco: Jossey-Bass Inc., Publishers, 1989.

Brandt, David. *Sacred cows make the best burgers*. New York: Warner Books, 1996.

Bridges, William *Managing transitions, making the most of change*. New York: Perseus Books. 1993.

Brookfield, Stephen D. *Developing critical thinkers*. San Francisco: Open University Press, 1987.

Bruegman, Randy R. *Exceeding customer expectations*. Upper Saddle River, NJ: Pearson Education, Inc., 2003.

Cohen, W. A. *The marketing plan*, 3rd ed. New York: John Wiley & Sons, Inc., 2001.

Coleman, Ron. "The Badge—Chief's Clipboard." *Fire Chief Magazine* 1991, pp. 24–26.

Coleman, W. A. *The marketing plan*, 3rd ed. New York: John Wiley & Sons, Inc., 2001.

Collins, James C. and Porras, Jerry I. "Building Your Company's Vision." *Harvard Business Review* September–October 1996, pp. 65–77.

Collins, Jim. "Level 5 Leadership: The Triumph of Humility and Fierce Resolve." *Harvard Business Review* January 2001, pp. 68–76.

Copeland, Lonnie. "A Multicultural Workforce." *Training Magazine* no date, pp. 50–56.

De Pree, Max. *Leadership without Power*. Holland, MI: Shepard Foundation, 1997.

Discovering the future: The business of paradigms. Lake Elmo, MN: ILI Press, 1988.

Dreyfuss, Joel. "Get Ready for a New Work Force." *Fortune* April 23, 1990, pp. 167–172, 181–182.

Fulghum, Robert. *All I really need to know I learned in kindergarten: Uncommon thoughts on common things*. New York: Ballantine Publishing Group, 1986.

Goleman, Daniel. *Emotional intelligence*. New York: Bantam Books, 1995.

Goleman, Daniel. "What Makes a Leader Humanized?" *Harvard Business Review* November–December 1998, pp. 94–102.

Great fires in America. Waukesha, WI: Country Beautiful Corporation.

Greenleaf, Robert. *Servant leadership*. Mahwah, NJ: Paulist Press, 1977.

Harari, Oren. *A leadership primer*. New York: American Management Association, December 1996, pp 34–38.

Harari, Oren. "Catapult Your Strategy Over Conventional Wisdom." *Management Review* October 1997, pp. 21–24.

"Hard Truths/Tough Choices." First Report on the National Commission for State and Local Public Service. The Nelson A. Rockefeller Institute of Government, 1993, Albany, NY.

Hashgen, Paul. "Firefighting in Colonial America." *Firehouse Magazine* September 1998, pp. 72–77.

Holman, Peggy and Devane, Tom. *The change handbook*. San Francisco: Berrett-Koehler Publishers, Inc., 1999.

Huy, Nguyen Quay. "In Praise of Middle Managers." *Harvard Business Review* September 2001, pp. 73–79.

Johnson, Spencer. *Who moved my cheese?* New York: G. P. Putnam's Sons, 1998.

Kapna, Robert E. and McCall, Morgan W., Jr. *Whatever it takes: The realities of managerial decision-making*, 2nd ed. Upper Saddle River, NJ: Prentice Hall, 2001.

Keirsey, David and Bates, Marilyn. *Please understand me*. Delmar, CA: Prometheus Nemesis Book Co., 1984.

Kiel, Joan M. "Reshaping Maslow's Hierarchy of Needs to Reflect Today's Educational and Managerial Philosophies." *Journal of Instructional Psychology* 1999, Vol. 26.

Kotter, John P. *Leading change*. Boston: Harvard Business School Press, 1996.

Kriegel, Robert and Patler, Louis. *If it ain't broke . . . break it*. New York: Warner Books, 1991.

Kuk, Michael. "The Maltese Cross and the Fire Service." *American Fire Journal* April 1998, p. 14.

Kuk, Michael. "The Legend of St. Florian, Patron Saint of Firefighters." *American Fire Journal* April 1999, pp. 55–58.

Linden, Russ. "The Human Side of Change." *International City/County Management Association* February 1997, pp. 1–13.

Lipka, Mitch. "Can You Trust Your Mover?" *Good Housekeeping* August 2003, p. 33.

Losyk, Bob. *Managing a changing workforce*. Davie, FL: Workplace Trends Publishing Co., 1995.

Lyones, Paul. *Fire in America*. Boston: National Fire Protection Association, 1976.

McCall, Morgan, Jr. and Robert Kaplan. *Whatever it takes: The realities of managerial decision making*, 2nd ed., Prentice Hall, Upper Saddle River, NJ, 1990.

McCurne, Jenny C. "The Elusive Thing Called Trust." *Management Review* July–August 1998, pp. 8–16.

McIntosh, Gary. *Three generations: Riding the wave of change in your church*. Grand Rapids, MI: Fleming H. Revell, 1995.

Merris, Jordan. *Have trust in partners*.

Mohrman, Allan M., et al. *Large-scale organizational change*. San Francisco: Jossey-Bass Inc., Publishers, 1989.

Musashi, Miyamoto. *A book of five rings: The classic guide to strategy*. Woodstock, NY: Overlook Press, 1974.

Nair, Keshavan. "The Core Value of Service." *Executive Intelligence Magazine* February 1996, pp. 11, 12.

Naisbitt, John and Aburdene, Patricia. *Ten new directions for the 1990s*. New York: Megatrends Ltd., 1990.

Osborne, David and Plastrick, Peter. *Banishing bureaucracy*. Reading, MA: Addison-Wesley Publishing Co., 1997.

Osborne, David and Plastrick, Peter. *The reinventor's fieldbook.* New York: Jossey-Bass Inc., Publishers, 2000.

Ostroff, Frank. *The horizontal organization.* New York: Oxford University Press.

O'Toole, James. *Leading change: Overcoming the ideology of comfort and the tyranny of customs.* San Francisco: Jossey–Bass Inc., Publishers, 1995.

Paulson, Terry L. *Paulson on change.* Glendale, CA: Griffin Publishing, 1995.

Peters, Tom and Waterman, Robert H. *In search of excellence.* New York: Warner Books, 1982.

Pfeffer, Jeffrey. "Providing Sustainable Competitive Advantage Through the Effective Management of People." *Academy of Management Executives* 1995, Vol. 9, no. 1, pp. 55–72.

Pound, Ron and Pritchett, Price. *The stress of organizational change.* Dallas, TX: Pritchett and Associates, 1998.

Powell, Colin. *My American dream.* New York: Random House, 1997.

Ricks, Thomas E. *Making the corps.* New York: Touchstone Books, 1998.

Roberts, Monty. *Horse sense for people.* New York: Viking Books, 2001.

Ruettiger, Daniel and Celizic, Mike. *Rudy's rules: Game plan for winning at life.* Waco, TX: WRS Publications, 1994.

Sher, Barbara. *How to get what you really want.* New York: Ballantine Books, 1979.

Smith, Doug. *Taking charge of change: 10 principles for managing people and performance.* New York: Perseus Publishing Co., 1997.

Tecker, Glenn H., Eide, Kermit M., and Frankel, Jean S. *Building a knowledge-based culture.* Washington, DC: American Society of Association Executives, 1997.

The organization of the future. New York: The Drucker Foundation, 1997.

"This Is Your Badge." *The Voice—Society of Fire Service Instruction.* December 1991, pp. 4–5.

Tichy, Noel M. and Devanna, Mary Anne. *The transformation leader.* New York: John Wiley & Sons, Inc., 1990.

Toffler, Alvin. *The adaptive corporation.* New York: McGraw-Hill Book Co., 1984.

Toffler, Alvin. *Powershift.* New York: Bantam Books, 1990.

Ulrich, Dave, Jick, Todd, and Kerr, Steve. *The boundaryless organization.* San Francisco: Jossey-Bass Inc., Publishers, 1995.

Wooden, John and Jamison, Steve. *A lifetime of observations on and off the court.* Chicago: Lincolnwood, 1997.

Zemke, Ron, Raines, Clair, and Filipczak, Bob. *Generations at work.* New York: Amacom Books, 2000.

INDEX